DISTRIBUTION RELIABILITY AND POWER QUALITY

DISTRIBUTION RELIABILITY AND POWER QUALITY

T. A. Short

EPRI Solutions, Inc.
Schenectady, NY

CRC Press is an imprint of the
Taylor & Francis Group, an **informa** business
A TAYLOR & FRANCIS BOOK

CRC Press
Taylor & Francis Group
6000 Broken Sound Parkway NW, Suite 300
Boca Raton, FL 33487-2742

ISBN-13: 978-0-8493-9575-8

Visit the Taylor & Francis Web site at
http://www.taylorandfrancis.com

and the CRC Press Web site at
http://www.crcpress.com

Library of Congress Cataloging-in-Publication Data

Short, T. A. (Tom A.), 1966-
Distribution reliability and power quality / by Thomas Allen Short.
p. cm.
Includes bibliographical references and index.
ISBN 0-8493-9575-5
1. Electric power distribution--Reliability. 2. Electric power system stability. 3. Electric power systems--Quality control. I. Title.

TK3091.S465 2005
333.793'2--dc22 2005050891

Dedication

To the future

To Jared

To Logan

Preface

In industrialized countries, distribution systems deliver electricity literally everywhere, taking power generated at many locations and delivering it to end users. Among generation, transmission, and distribution — the big three components of the electricity infrastructure — the distribution system gets the least attention. Yet, it is often the most critical component in terms of its effect on reliability and quality of service, cost of electricity, and aesthetic (mainly visual) impacts on society.

Like much of the electric utility industry, several political, economic, and technical changes are pressuring the way distribution systems are built and operated. Deregulation has increased pressures on electric power utilities to cut costs and has focused emphasis on reliability and quality of electric service. The great fear of deregulation is that service will suffer because of cost cutting. Regulators and utility consumers are paying considerable attention to reliability and quality. Customers are pressing for lower costs and better reliability and power quality. The performance of the distribution system determines greater than 90% of the reliability of service to customers (the high-voltage transmission and generation system determine the rest). If performance is increased, it will have to be done on the distribution system.

This book is a spin-off from the *Electric Power Distribution Handbook* (2004) that includes the portions of that handbook that target reliability and power quality. The main focus of this book is on the most critical quality-of-supply issues: short- and long-duration interruptions and voltage sags. The book provides ways to analyze these and limit the impact on end-use customers. The book contains a new chapter that was not in the 2004 handbook entitled "Reliability and Power Quality Improvement Programs" that was developed with Lee Taylor of Duke Power. This includes practical advice on developing programs to improve power quality and reliability.

I hope you find useful information in this book. If it's not in here, hopefully, one of the many bibliographic references will lead you to what you're looking for. Please feel free to email me feedback on this book including errors, comments, opinions, or new sources of information. Also, if you need my help with any interesting consulting or research opportunities, I'd love to hear from you.

Tom Short
EPRI Solutions, Inc.
Schenectady, NY
t.short@ieee.org

Acknowledgments

First and foremost, I would like to thank my wife Kristin; thank you for your strength, thank you for your help, thank you for your patience, and thank you for your love. My play buddies, Logan and Jared, energized me and made me laugh. My family was a source of inspiration. I would like to thank my parents, Bob and Sandy, for their influence and education over the years.

EPRI Solutions, Inc. (formerly EPRI PEAC) provided a great deal of support on this project. I would like to recognize the reviews, ideas, and support of Phil Barker and Dave Crudele in Schenectady, New York, and also Arshad Mansoor, Mike Howard, Charles Perry, Arindam Maitra, and the rest of the energetic crew in Knoxville, Tennessee.

Many other people reviewed portions of the draft and provided input and suggestions, including Dave Smith (Power Technologies, Inc.), Dan Ward (Dominion Virginia Power), Jim Stewart (Consultant, Scotia, NY), Conrad St. Pierre (Electric Power Consultants), Karl Fender (Cooper Power Systems), John Leach (Hi-Tech Fuses, Inc.), and Rusty Bascom (Power Delivery Consultants, LLC).

Thanks to Power Technologies, Inc. for opportunities and mentoring during my early career with the help of several talented, helpful engineers, including Jim Burke, Phil Barker, Dave Smith, Jim Stewart, and John Anderson. Over the years, several clients have also educated me in many ways; two that stand out include Ron Ammon (Keyspan, retired) and Clay Burns (National Grid).

EPRI has been supportive of this project, including a review by Luther Dow. The company has also sponsored a number of interesting distribution research projects with which I have been fortunate enough to be involved and has allowed me to share some of those efforts here.

Lee Taylor of Duke Power worked with me to develop the new chapter on programs to improve reliability and power quality (Chapter 5). Lee and Duke Power pioneered the strategy of identifying and removing sources of faults on the Duke Power system. By doing this, Duke Power is able to maintain very respectable reliability numbers despite the facts that its service territory has regular severe weather and that Duke Power is mainly an overhead utility with predominantly all-radial systems. Lee's experience using reliability databases and developing programs to improve reliability significantly improved this book. Duke Power also provided a number of educational photographs that serve as useful examples.

As a side note, I would like to recognize the efforts of linemen in the electric power industry. These folks do the real work of building the lines and keeping the power on. As a tribute to them, a trailer at the end of each chapter reveals a bit of the lineman's character and point of view.

About the Author

Tom Short has spent most of his career working on projects helping utilities improve their reliability and power quality. He performed lightning protection, reliability, and power quality studies for many utility distribution systems while at Power Technologies, Inc. from 1990 through 2000. He has done extensive digital simulations of T&D systems using various software tools, including EMTP to model lightning surges on overhead lines and underground cables, distributed generators, ferroresonance, faults and voltage sags, and capacitor switching. Since joining EPRI PEAC in 2000 (now EPRI Solutions, Inc.), Mr. Short has led a variety of distribution research projects for the company, including a capacitor reliability initiative, a power quality handbook for distribution companies, a distributed generation workbook, and a series of projects directed at improving distribution reliability and power quality.

As chair of the IEEE Working Group on the Lightning Performance of Distribution Lines, Mr. Short led the development of IEEE Std. 1410-1997, "Improving the Lightning Performance of Electric Power Overhead Distribution Lines." He was awarded the 2002 Technical Committee Distinguished Service Award by the IEEE Power Engineering Society for this effort.

Mr. Short has also performed a variety of other studies including railroad impacts on a utility (flicker, unbalance, and harmonics), load flow analysis, capacitor application, loss evaluation, and conductor burndown. He has taught courses on reliability, power quality, lightning protection, overcurrent protection, harmonics, voltage regulation, capacitor application, and distribution planning.

Mr. Short developed the Rpad engineering analysis interface (www.Rpad.org) that EPRI Solutions, Inc. is using to offer engineering, information, mapping, and database solutions to electric utilities. Rpad is an interactive, Web-based analysis program. Its pages are interactive workbook-type sheets based on R, an open-source implementation of the S language (used to make many of the graphs in this book). Rpad is an analysis package, a Web-page designer, and a gui designer, all wrapped in one. This package makes it easy to develop powerful data-analysis applications that can be easily shared on a company intranet.

Mr. Short graduated with a master's degree in electrical engineering from Montana State University in 1990 after receiving a bachelor's degree in 1988.

Credits

Figure 3.19 is reprinted with permission from IEEE Std. 141-1993. *IEEE Recommended Practice for Electric Power Distribution for Industrial Plants.* Copyright 1994 by IEEE.

Tables 3.1 to 3.3 are reprinted with permission from IEEE Std. 519-1992. *IEEE Recommended Practices and Requirements for Harmonic Control in Electrical Power Systems.* Copyright 1993 by IEEE.

Table 4.9 is reprinted with permission from IEEE Std. 493-1997. *IEEE Recommended Practice for the Design of Reliable Industrial and Commercial Power Systems (Gold Book).* Copyright 1998 by IEEE.

Contents

1 Reliability **1**
1.1 Reliability Indices 1
1.1.1 Customer-Based Indices 1
1.1.2 Load-Based Indices 5
1.2 Storms and Weather 6
1.3 Variables Affecting Reliability Indices 10
1.3.1 Circuit Exposure and Load Density 10
1.3.2 Supply Configuration 11
1.3.3 Voltage 12
1.3.4 Long-Term Reliability Trends 13
1.4 Modeling Radial Distribution Circuits 15
1.5 Parallel Distribution Systems 17
1.6 Improving Reliability 21
1.6.1 Identify and Target Fault Causes 22
1.6.2 Identify and Target Circuits 23
1.6.3 Switching and Protection Equipment 23
1.6.4 Automation 27
1.6.5 Maintenance and Inspections 29
1.6.6 Restoration 31
1.6.7 Fault Reduction 34
1.7 Interruption Costs 34
References 36

2 Voltage Sags and Momentary Interruptions **39**
2.1 Location 40
2.2 Momentary Interruptions 42
2.3 Voltage Sags 45
2.3.1 Effect of Phases 51
2.3.2 Load Response 51
2.3.3 Analysis of Voltage Sags 53
2.4 Characterizing Sags and Momentaries 54
2.4.1 Industry Standards 54
2.4.2 Characterization Details 56
2.5 Occurrences of Voltage Sags 57
2.5.1 Site Power Quality Variations 59
2.5.2 Transmission-Level Power Quality 62
2.6 Correlations of Sags and Momentaries 62
2.7 Factors That Influence Sag and Momentary Rates 63
2.7.1 Location 64

2.7.2 Load Density 66
2.7.3 Voltage Class....67
2.7.4 Comparison and Ranking of Factors....67
2.8 Prediction of Quality Indicators Based on Site Characteristics....69
2.9 Equipment Sensitivities....71
2.9.1 Computers and Electronic Power Supplies....71
2.9.2 Industrial Processes and Equipment....75
2.9.2.1 Relays and Contactors....76
2.9.2.2 Adjustable-Speed Drives....79
2.9.2.3 Programmable-Logic Controllers....81
2.9.3 Residential Equipment....81
2.10 Solution Options....81
2.10.1 Utility Options for Momentary Interruptions....81
2.10.2 Utility Options for Voltage Sags....84
2.10.2.1 Raising the Nominal Voltage....85
2.10.2.2 Line Reactors....85
2.10.2.3 Neutral Reactors....86
2.10.2.4 Current-Limiting Fuses....87
2.10.3 Utility Options with Nontraditional Equipment....88
2.10.3.1 Fast Transfer Switches....88
2.10.3.2 DVRs and Other Custom-Power Devices....89
2.10.4 Customer/Equipment Solutions....91
2.11 Power Quality Monitoring....92
References....94

3 Other Power Quality Issues....99
3.1 Overvoltages and Customer Equipment Failures....99
3.1.1 Secondary/Facility Grounding....101
3.1.2 Reclose Transients....103
3.2 Switching Surges....104
3.2.1 Voltage Magnification....108
3.2.2 Tripping of Adjustable-Speed Drives....110
3.2.3 Prevention of Capacitor Transients....111
3.3 Harmonics....112
3.3.1 Resonances....119
3.3.2 Telephone Interference....121
3.4 Flicker....125
3.4.1 Flicker Solutions....133
3.4.1.1 Load Changes....133
3.4.1.2 Series Capacitor....135
3.4.1.3 Static Var Compensator....138
3.4.1.4 Other Solutions....139
3.5 Voltage Unbalance....140
References....143

4 Faults **147**
4.1 General Fault Characteristics 147
4.2 Fault Calculations 153
4.2.1 Transformer Connections 158
4.2.2 Fault Profiles 159
4.2.3 Effect of *X/R* Ratio 162
4.2.4 Secondary Faults 165
4.2.5 Primary-to-Secondary Faults 167
4.2.6 Underbuilt Fault to a Transmission Circuit 171
4.2.7 Fault Location Calculations 174
4.3 Limiting Fault Currents 178
4.4 Arc Characteristics 179
4.5 High-Impedance Faults 185
4.6 External Fault Causes 189
4.6.1 Trees 189
4.6.2 Weather and Lightning 196
4.6.3 Animals 197
4.6.4 Other External Causes 198
4.7 Equipment Faults 199
4.8 Faults in Equipment 200
References 204

5 Reliability and Power Quality Improvement Programs **209**
5.1 Improvements in Protection Practices 209
5.1.1 Fusing 210
5.1.2 Fuse Saving vs. Fuse Blowing 211
5.1.3 Reclosing Practices. 215
5.1.4 Single-Phase Protective Devices 218
5.1.4.1 Ferroresonance 219
5.1.4.2 Backfeeds 219
5.1.4.3 Single-Phasing Impacts on Motors 220
5.1.4.4 Single-Phase Trip, Three-Phase Lockout 221
5.1.5 Improving Coordination 221
5.1.6 Locating Sectionalizing Equipment 222
5.2 Fault Sources 226
5.2.1 Trees 228
5.2.2 Lightning 233
5.2.3 Animals 240
5.2.4 Cable and Equipment Failures 243
5.3 Programs to Reduce Fault Rates 246
5.4 Outage Follow-Ups 247
5.5 Problem-Circuit Audits 249
5.6 Construction Upgrade Programs 250
5.7 Using Outage Databases 252
References 257

1

Reliability

Power outages disrupt more businesses than any other factor (see Figure 1.1). I lose two hours of work on the computer; Jane Doe gets stuck in an elevator; Intel loses a million dollars worth of computer chips; a refinery flames out, stopping production and spewing pollution into the air. End users expect good reliability, and expectations keep rising. Interruptions and voltage sags cause most disruptions. In this chapter we study "sustained" interruptions, long-duration interruptions generally defined as lasting longer than 1 to 5 min. We investigate momentary interruptions and voltage sags in the next chapter. Reliability statistics, based on long-duration interruptions, are the primary benchmark used by utilities and regulators to identify service quality. Faults on the distribution system cause most long-duration interruptions; a fuse, breaker, recloser, or sectionalizer locks out the faulted section.

Many utilities use reliability indices to track the performance of the utility or a region or a circuit. Regulators require most investor-owned utilities to report their reliability indices. The regulatory trend is moving to performance-based rates where performance is penalized or rewarded based on quantification by reliability indices. Some utilities also pay bonuses to managers or others based in part on indices. Some commercial and industrial customers ask utilities for their reliability indices when locating a facility.

1.1 Reliability Indices

1.1.1 Customer-Based Indices

Utilities most commonly use two indices, SAIFI and SAIDI, to benchmark reliability. These characterize the frequency and duration of interruptions during the reporting period (usually years) (IEEE Std. 1366-2000).

SAIFI, System average interruption frequency index

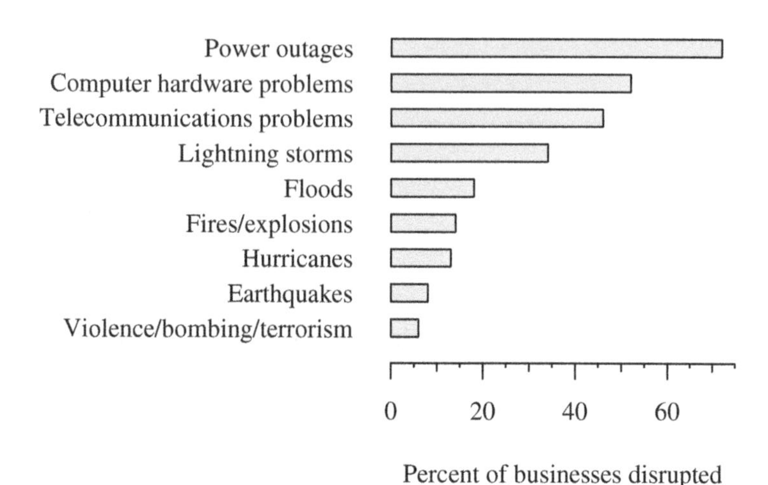

FIGURE 1.1
Percent of U.S. businesses disrupted by the given problem. (Data from [Rodentis, 1999].)

$$\text{SAIFI} = \frac{\text{Total number of customer interruptions}}{\text{Total number of customers served}}$$

Typically, a utility's customers average between one and two sustained interruptions per year. SAIFI is also the average failure rate, which is often labeled λ. Another useful measure is the mean time between failure (MTBF), which is the reciprocal of the failure rate: MTBF in years = $1/\lambda$.

SAIDI, System average interruption duration frequency index

$$\text{SAIDI} = \frac{\text{Sum of all customer interruption durations}}{\text{Total number of customers served}}$$

SAIDI quantifies the average total duration of interruptions. SAIDI is cited in units of hours or minutes per year. Other common names for SAIDI are CMI and CMO, standing for customer minutes of interruption or outage.

SAIFI and SAIDI are the most-used pair out of many reliability indices, which look like a wash of acronyms — most importantly, *D* for duration and *F* for frequency. Another related index is CAIDI:

CAIDI, Customer average interruption duration frequency index

$$\text{CAIDI} = \frac{\text{SAIDI}}{\text{SAIFI}} = \frac{\text{Sum of all customer interruption durations}}{\text{Total number of customer interruptions}}$$

CAIDI is the "apparent" repair time (from the customers' perspective). It is generally much shorter than the actual repair time because utilities nor-

TABLE 1.1

Reliability Indices Found by Industry Surveys

	SAIFI, No. of Interruptions/Year			SAIDI, h of Interruption/Year		
	25%	**50%**	**75%**	**25%**	**50%**	**75%**
IEEE Std. 1366-2000	0.90	1.10	1.45	0.89	1.50	2.30
EEI (1999) [excludes storms]	0.92	1.32	1.71	1.16	1.74	2.23
EEI (1999) [with storms]	1.11	1.33	2.15	1.36	3.00	4.38
CEA (2001) [with storms]	1.03	1.95	3.16	0.73	2.26	3.28
PA Consulting (2001) [with storms]				1.55	3.05	8.35
IP&L Large City Comparison (Indianapolis Power & Light, 2000)	0.72	0.95	1.15	1.02	1.64	2.41

Note: 25%, 50%, and 75% represent the lower quartile, the median, and the upper quartile of utilities surveyed.

mally sectionalize circuits to reenergize as many customers as possible before crews fix the actual damage.

Also used in many other industries, the availability is quantified as

ASAI, Average service availability index

$$\text{ASAI} = \frac{\text{Customer hours service availability}}{\text{Customer hours service demanded}}$$

We can find ASIFI from SAIDI specified in hours as

$$\text{ASAI} = \frac{8760 - \text{SAIDI}}{8760}$$

(Use 8784 h/year for a leap year.)

Survey results for SAIFI and SAIDI are shown in Table 1.1. Figure 1.2 shows the distribution of utility indices from the CEA survey. Much of this reliability index data is from Short (2002). Utility indices vary widely because of many differing factors, mainly

- Weather
- Physical environment (mainly the amount of tree coverage)
- Load density
- Distribution voltage
- Age
- Percent underground
- Methods of recording interruptions

Within a utility, performance of circuits varies widely for many of the same reasons causing the spread in utility indices: circuits have different lengths

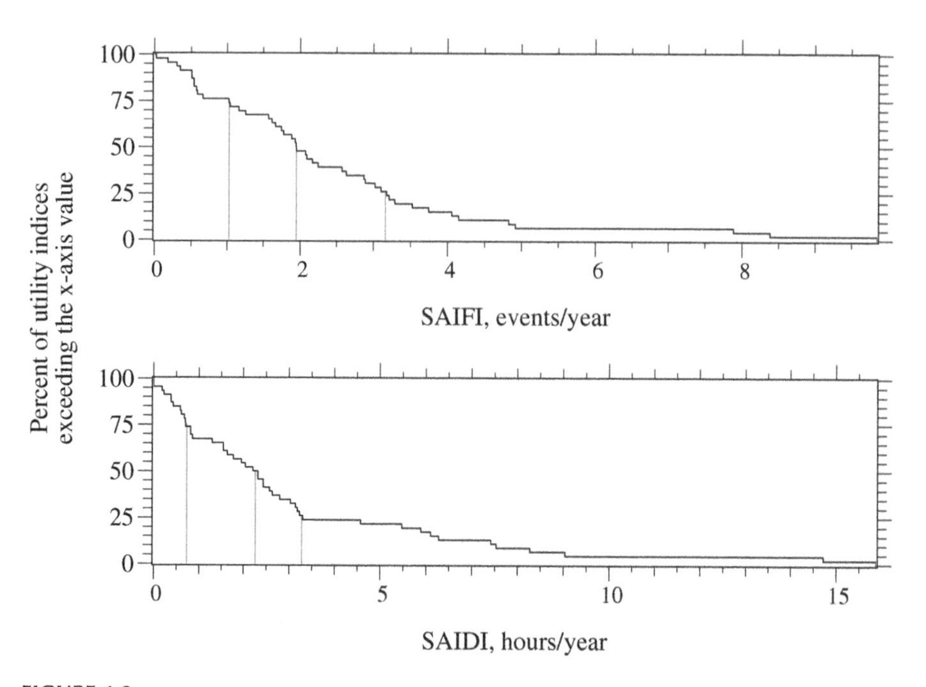

FIGURE 1.2
Distribution of utility indices in Canada (CEA survey, 36 utilities, two-year average). (Data from [CEA, 2000].)

necessary to feed different areas of load density, some are older than others, and some areas may have less tree coverage. Figure 1.3 shows the spread of reliability on individual feeders at two utilities for two years worth of data. Even though these two utilities are within the same state, SAIFI differs dramatically.

Customer reliability is not normally distributed. A skewed distribution such as the log-normal distribution is more appropriate and has been used in several reliability applications (Brown and Burke, 2000; Christie, 2002). A log-normal distribution is appropriate for data that is bounded on the lower side by zero. The skewed distribution has several ramifications:

- The average is higher than the median. The median is a better representation of the "typical" customer.
- Poor performing customers and circuits dominate the indices (which are averages).
- Storms and other outliers easily skew the indices.

Realize that SAIFI and SAIDI are weighted performance indices. They stress the performance of the worst-performing circuits and the performance during storms. SAIFI and SAIDI are not necessarily good indicators of the typical performance that customers have.

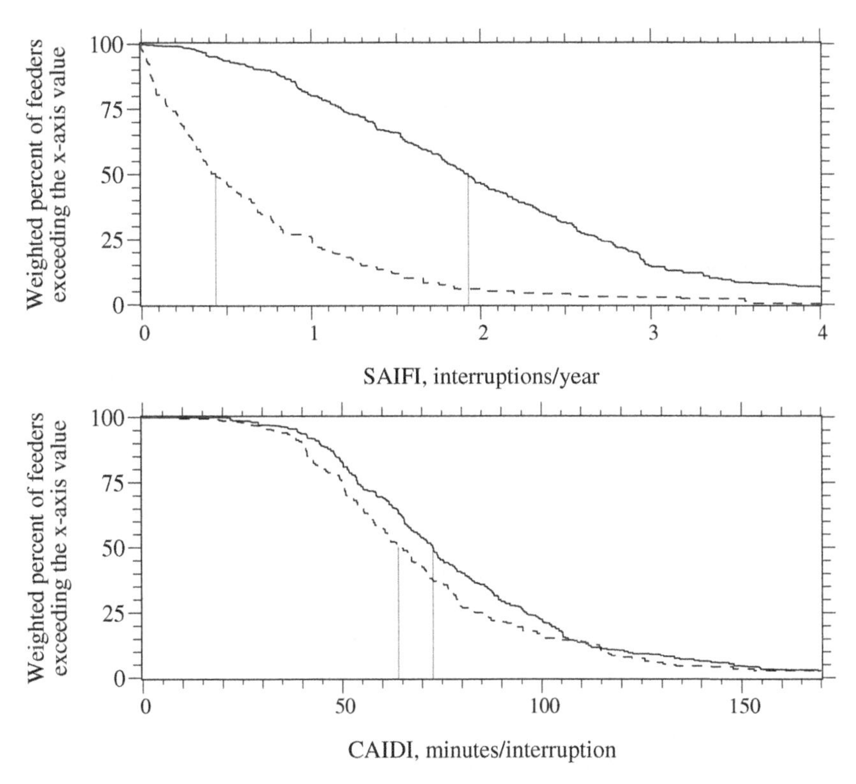

FIGURE 1.3
Reliability indices by feeder for two utilities. Forced events only—major events, scheduled events, and outside causes (substation or transmission) are excluded. The total SAIFI including all events was 0.79 for Utility A and 3.4 for Utility B.

1.1.2 Load-Based Indices

Residential customers dominate SAIFI and SAIDI since these indices treat each customer the same. Even though residential customers make up 80% of a typical utility's customer count, they may only have 40% of the utility's load. To more fairly weight larger customers, load-based indices are available; the equivalent of SAIFI and SAIDI, but scaled by load, are ASIFI and ASIDI:

ASIFI, Average system interruption frequency index

$$\text{ASIFI} = \frac{\text{Connected kVA interrupted}}{\text{Total connected kVA served}}(\text{Average number of interruptions})$$

ASIDI, Average system interruption frequency index

$$\text{ASIDI} = \frac{\text{Connected kVA interruption duration}}{\text{Total connected kVA served}}$$

Fewer than 8% of utilities track ASIFI and ASIDI, mainly since they are hard to track (knowing load interrupted is more difficult than knowing number of customers interrupted). Utilities also feel that commercial and industrial customers have enough clout that their problems are given due attention.

1.2 Storms and Weather

Much of the reliability data reported to regulators excludes major storm or major event interruptions. There are pros and cons to excluding storm interruptions. The argument for excluding storms is that storm interruptions significantly alter the duration indices to the extent that restoration performance dominates the index. Further, a utility's performance during storms does not necessarily represent the true performance of the distribution system. Including storms also adds considerable year-to-year variation in results. On the other hand, from the customer point of view an interruption is still an interruption. Also, the performance of a distribution system is reflected in the storm performance; for example, if a utility does more tree trimming and puts more circuits underground, their circuits will have fewer interruptions when a storm hits.

If storms are not excluded, the numbers go up as shown in an EEI survey in Figure 1.4. The interruption duration (CAIDI) and the average total interruption time (SAIDI) increase the most if storm data is included. Storms only moderately impact SAIFI. During storms, crew resources are fully used. Downed trees and wires plus traffic makes even getting to faults difficult. Add difficult working conditions, and it is easy to understand the great increase in repair times.

During severe storms, foreign crews, crews from other service territories, and general mayhem add large roadblocks preventing utilities from keeping records needed for tracking indices. Expedience rules — should I get the lights back on or do paperwork?

Utilities use various methods to classify storms. The two common categories are

- *Statistical method* — A common definition is 10% of customers affected within an operating area.
- *Weather based definition* — Common definitions are "interruptions caused by storms named by the national weather service" and "interruptions caused during storms that lead to a declaration of a state of emergency."

Some utilities exclude other interruptions including those scheduled or those from other parts of the utility system (normally substation or trans-

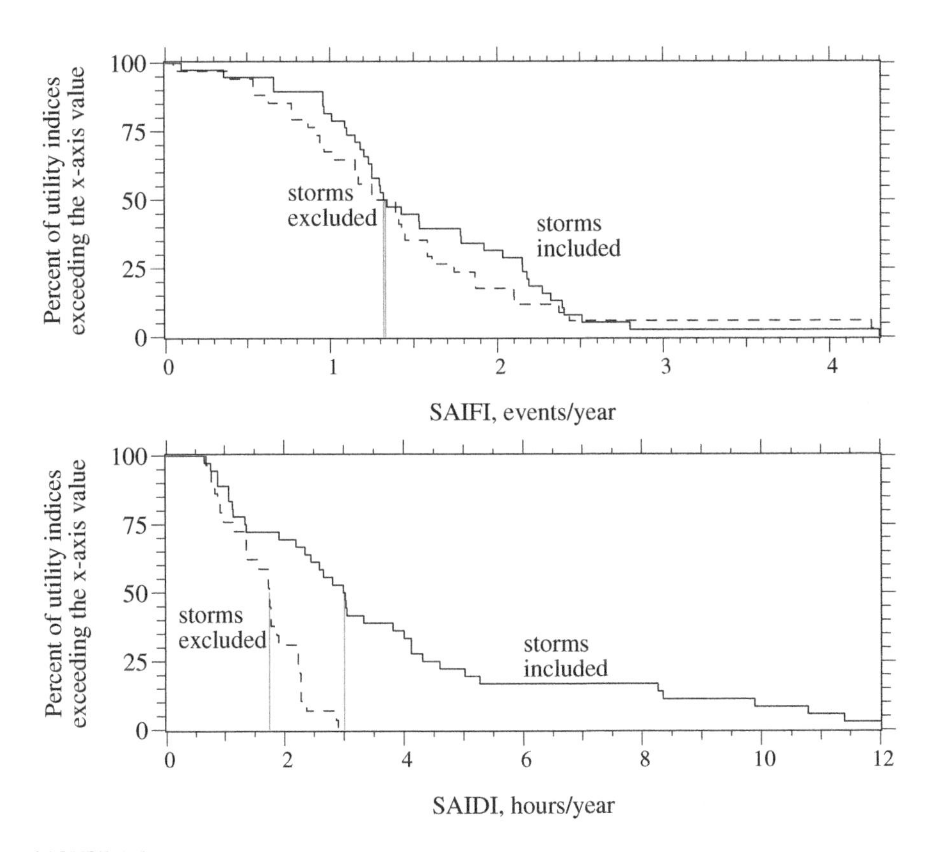

FIGURE 1.4
Distribution of utility indices with and without excluding storms. (Data from [EEI, 1999].)

mission-caused interruptions). Both are done for the same reasons as storm exclusions: neither scheduled interruptions nor transmission-caused interruptions reflect the normal operating performance of the distribution system.

The IEEE working group appears to favor a statistical approach to classifying major events (Christie, 2002). An argument against this approach is that major substation or transmission outages can be "major events" and get excluded from indices. From the customer point of view, major event or no major event, an interruption is still a loss of production or a spoiled inventory or a loss of productivity or a missed football game. For this reason, some regulators hesitate to allow exclusions.

During storm days, the interruption durations increase exponentially. Figure 1.5 shows probability distributions of the daily SAIDI based on data from four utilities. The plot is on a log-normal scale: the x-axis shows SAIDI for each day on a log scale, and the y-axis shows the probability on a normal-distribution scale. On this plot, data with a log-normal distribution comes out as a straight line. Most of the utility data fits a log-normal distribution. But, two of the utilities are even more skewed than a log-normal distribution indicates — at these utilities, storm days have even more customer-minutes

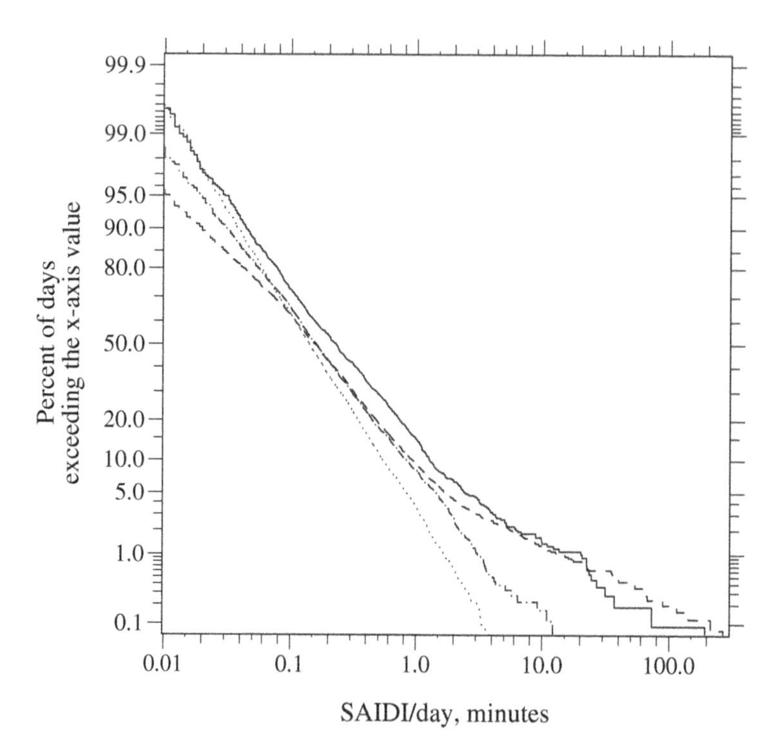

FIGURE 1.5
SAIDI per day probability distributions for four utilities.

interrupted. At one of these utilities, 0.2% of the days contributed 40% of the SAIDI index (over a 7-year period). During the worst year at this utility, 70% of SAIDI for that year happened because of three storms (impacting 5 days total). If we have one day with a SAIDI of over 100 min (the value for a whole year at some utilities), it is going to be a long year. Figure 1.6 again emphasizes how skewed the probability distribution of reliability data is. The average is much higher than the median, and the extreme days heavily influence the average.

Weather, even if it does not reach the level of "major event," plays a major role in reliability. Weather varies considerably from year to year — these weather variations directly affect reliability indices. Bad lightning years or excessively hot years worsen the indices.

Monitoring by Ontario Hydro Technologies in Ontario, Canada, gives some insight into storm durations and failure rates (CEA 160 D 597, 1998). In a mild to moderate lightning area, 20% of the interruptions occurred during storm periods and 15% in the 24 h following a storm. The study area had an average of 25 storm days per year and 73 storm hours per year. Therefore, about 35% of outages occur in 7% of the time in a year (and 20% in 0.8% of the time). The study found an interruption rate during storms of 10 to 20 times the non-storm rate.

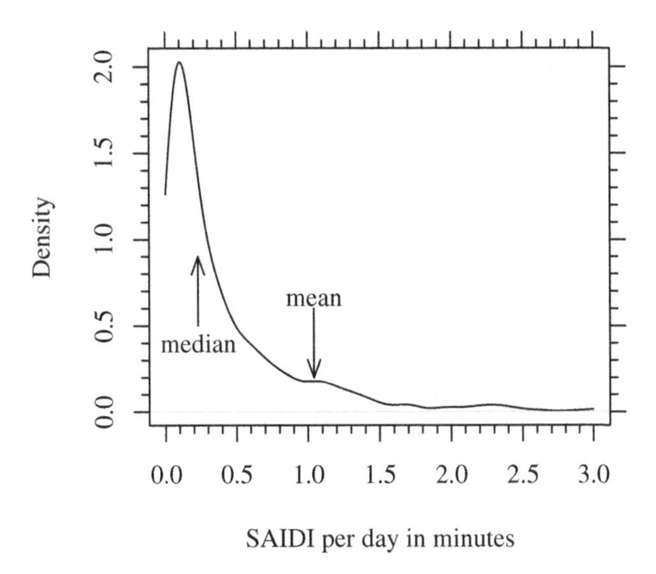

FIGURE 1.6
SAIDI per day probability density at one utility.

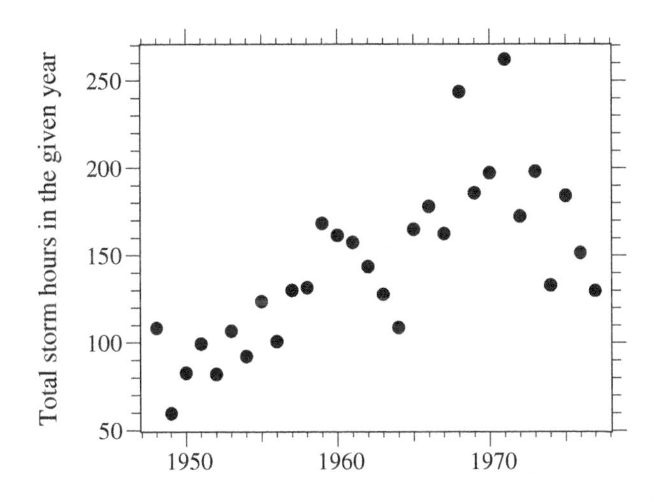

FIGURE 1.7
Thunderstorm duration by year for Tampa, Florida. (Data from [MacGorman et al., 1984].)

Faults and interruptions have significant year-to-year variation because weather conditions vary significantly. Just as severe storm patterns vary, normal storm frequencies and durations vary. Consider the thunderstorm duration plot in Figure 1.7. Over this 30-year period in Tampa, FL (a very high lightning area), some years had less than 80 h of storms, and a couple of years had more than 240 h. These are not "severe events," just variations in the normal weather patterns. These storm variations translate into variations in the number of faults and in the reliability indices. Even in areas with lower storm activity, significant variation is possible. Consider these variations if reliability "baselines" are going to be set for performance-based rates. Wind, icing, and temperature extremes all have significant year-to-year variations that directly impact reliability indices. Watch out for a few years of consistent weather; if the data from 1950–1955 of Figure 1.7 were used for performance-based rates, we would be in trouble in following years.

The first step in quantifying the effect of weather on interruptions is to track weather statistics along with interruption statistics. Lightning, wind, temperature, and other important weather statistics are available from national weather services as well as private groups, and many statistics have long historical records. Correlations between weather statistics and interruptions can help quantify the variations. Brown et al. (1997) show an example for a feeder in Washington state where wind-dependent failures were analyzed. For this case, they found 0.0065 failures/mi/year/mph of wind speed.

After correlating interruptions with weather data, we can extrapolate how much reliability indices could vary using historical weather data. One could even use weather statistics to come up with a normalized interruption index that tried to smooth out the weather variations.

1.3 Variables Affecting Reliability Indices

1.3.1 Circuit Exposure and Load Density

Longer circuits lead to more interruptions. This is difficult to avoid on normal radial circuits, even though we can somewhat compensate by adding reclosers, fuses, extra switching points, or automation. Most of the change is in SAIFI; the interruption duration (CAIDI) is less dependent on load circuit lengths. Figure 1.8 shows the effect on SAIFI at one utility in the southwest U.S.

It is easier to provide higher reliability in urban areas: circuit lengths are shorter, and more reliable distribution systems (such as a grid network) are more economical. The Indianapolis Power and Light survey results shown in Figure 1.9 only included performance of utilities in large cities. As expected, the urban results are better than other general utility surveys. Another comparison is shown in Figure 1.10 — in all states, utilities with higher load densities tend to have better SAIFIs.

Figure 1.11 and Figure 1.12 show reliability for different distribution services in several Commonwealth countries. The delineations used for this comparison for Victoria are

- Central business district: used map boundaries
- Urban: greater than 0.48 MVA/mi (0.3 MVA/km)

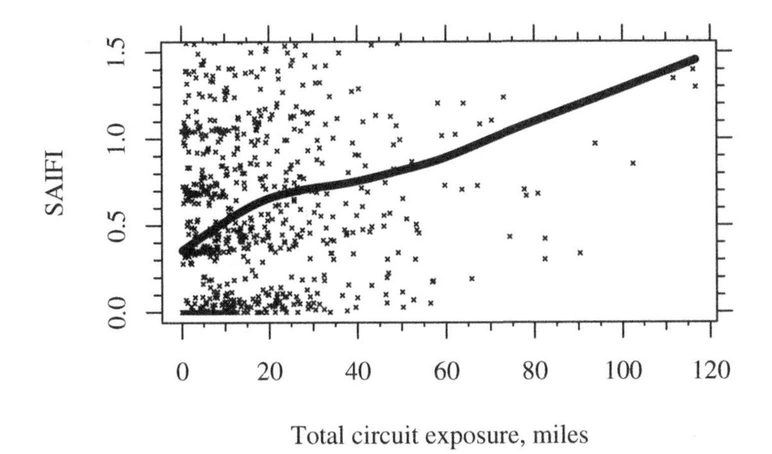

FIGURE 1.8
Effect of circuit length on SAIFI for one utility in the southwest U.S.

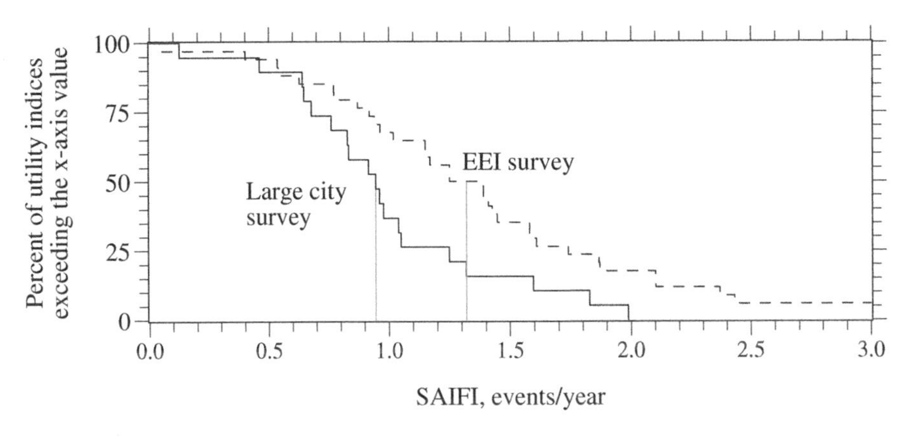

FIGURE 1.9
Comparison of the Indianapolis Power & Light Large City Survey of SAIFI to the general EEI survey results (with storms excluded). (Data from [Indianapolis Power & Light, 2000].)

- Short rural: less than 124 mi (200 km)
- Long rural: greater than 124 mi (200 km)

1.3.2 Supply Configuration

The distribution supply greatly impacts reliability. Long radial circuits provide the poorest service; grid networks provide exceptionally reliable service. Table 1.2 gives estimates of the reliability of several common distribution supply types developed by New York City's Consolidated Edison. Massive redundancy for grid and spot networks leads to fantastic reliability — 50 plus years between interruptions. Note that the interruption duration

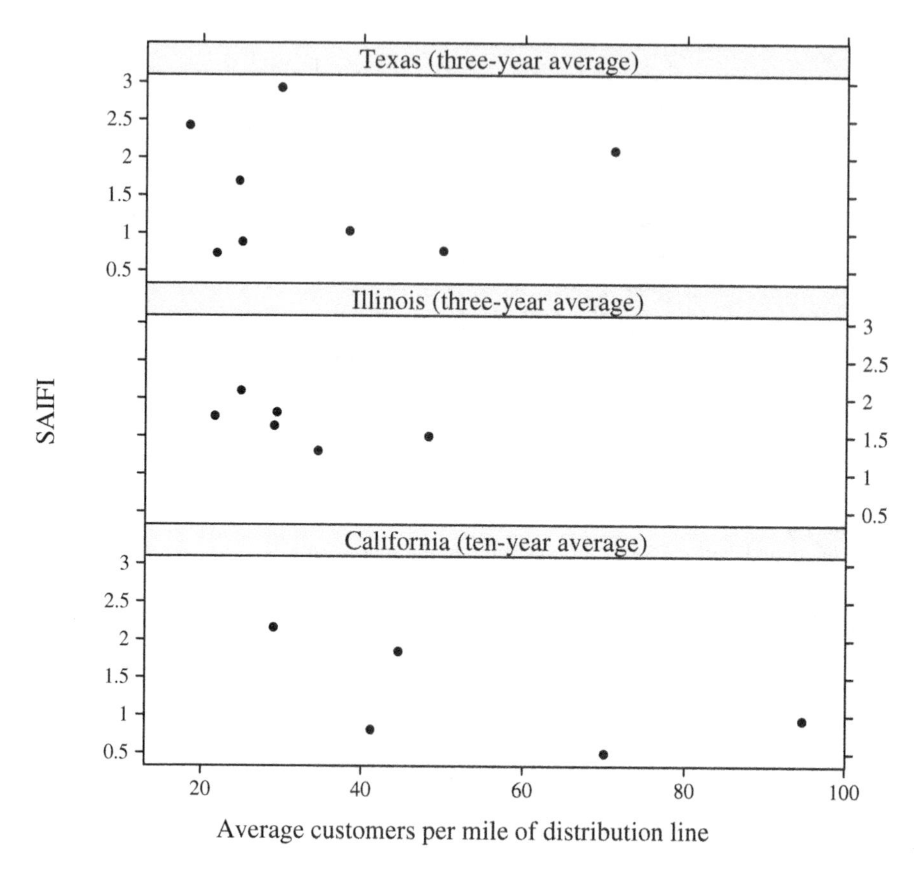

FIGURE 1.10
Effect of customer density on SAIFI.

(CAIDI) increases for the more urban configurations. Being underground and dealing with traffic increase the time for repairs.

1.3.3 Voltage

Higher primary voltages tend to be more unreliable, mainly because of longer lines. Figure 1.13 shows an example for one utility that is typical of many utilities: higher voltage circuits have more interruptions.

On higher-voltage primary circuits, we need to make more of an effort to achieve the same reliability as for lower voltage circuits: more reclosers, more sectionalizing switches, more tree trimming, and so forth. With the ability to build much longer lines and serve more customers, it is difficult to overcome the increased exposure. Keeping reliability in mind when planning higher-voltage systems helps. On higher-voltage circuits, wider is better than longer. Burke's analysis (1994) of the service length and width for a generalized feeder shows that for the best reliability, higher-voltage circuits should

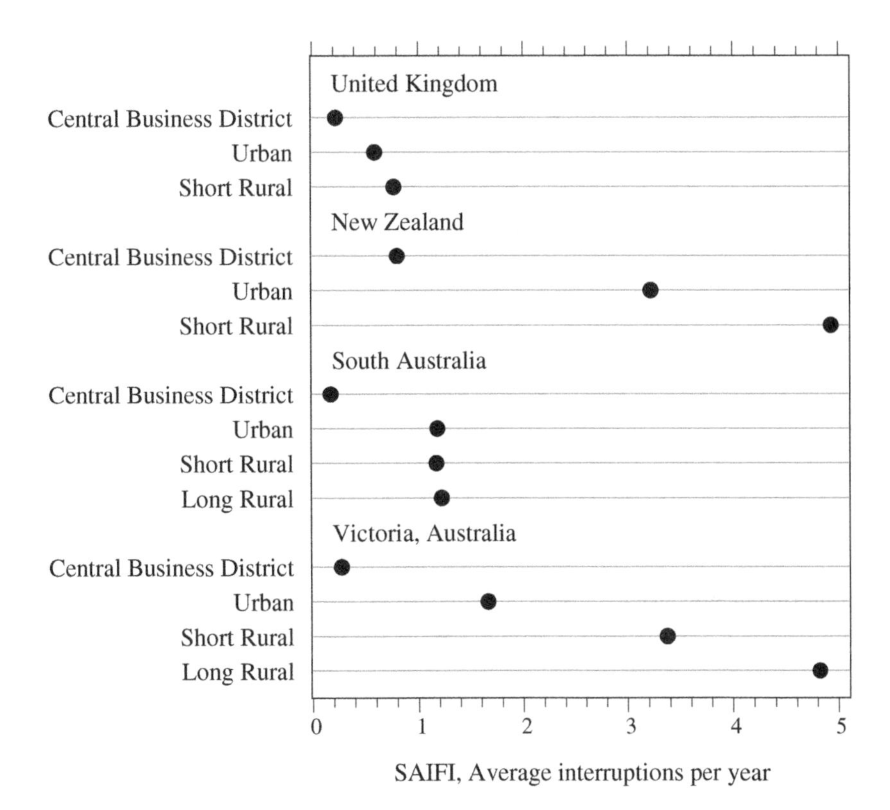

FIGURE 1.11
Comparison of SAIFI by load density for several former British Empire colonies. (Data from [Coulter, 1999].)

be longer and wider, not just longer (see Table 1.3). Usually, higher-voltage circuits are just made longer, which leads to poor reliability. Having a long skinny main feeder with short taps off of the mainline results in poor reliability performance.

1.3.4 Long-Term Reliability Trends

Utilities rarely have very long-term data covering decades. The Canadian Electrical Association has tracked reliability data for many years. Figure 1.14 shows SAIDI over a 40-year period for Canada. Significant variation exists from year to year. Part of this is due to the changing nature of the survey (the utility base was not consistent for the whole time period). Much of the variation is due to weather, even though the survey covers a huge geographic area (we expect more variations for smaller geographic areas). The data includes storms. Extreme years stand out. The worst year was 1998, which was dominated by the ice storm that hit Ontario and Quebec. Over 1.6 million customers lost power; Hydro Quebec's SAIDI for the year was almost 42 h when it is normally less than 4 h.

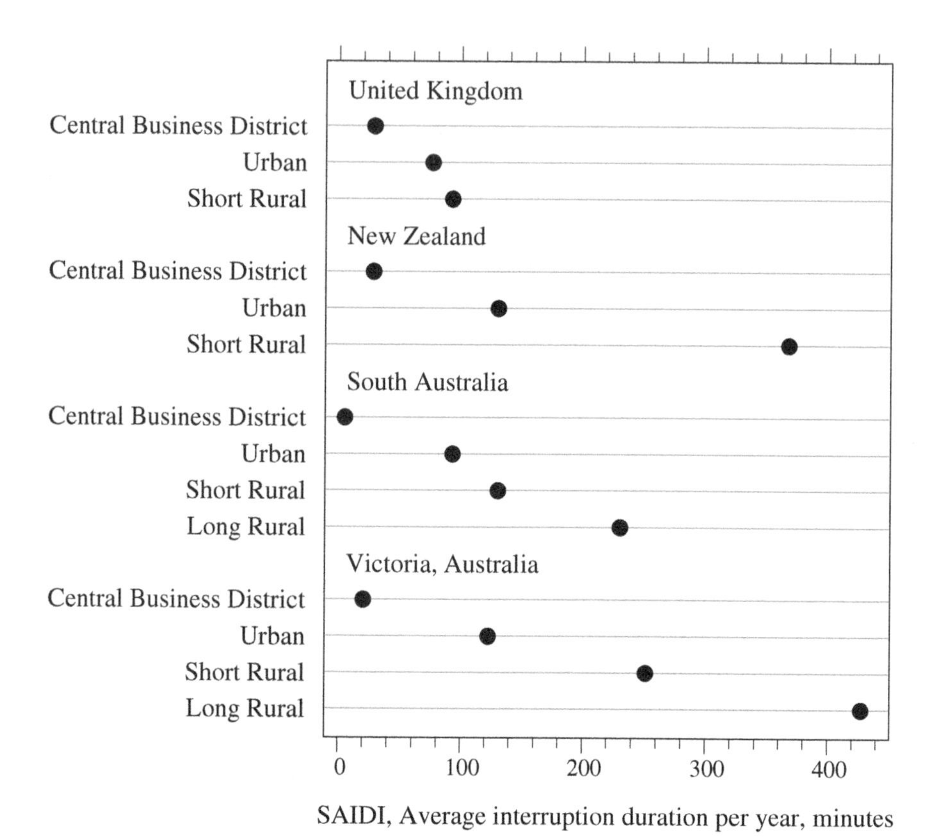

FIGURE 1.12
Comparison of SAIDI by load density for several former British Empire colonies. (Data from [Coulter, 1999].)

TABLE 1.2

Comparison of the Reliability of Different Distribution Configurations

	SAIFI Interruptions/Year	CAIDI min/Interruption	MAIFI Momentary Interruptions/Year
Simple radial	0.3 to 1.3	90	5 to 10
Primary auto-loop	0.4 to 0.7	65	10 to 15
Underground residential	0.4 to 0.7	60	4 to 8
Primary selective	0.1 to 0.5	180	4 to 8
Secondary selective	0.1 to 0.5	180	2 to 4
Spot network	0.02 to 0.1	180	0 to 1
Grid network	0.005 to 0.02	135	0

Source: Settembrini, R. C., Fisher, J. R., and Hudak, N. E., "Reliability and Quality Comparisons of Electric Power Distribution Systems," IEEE Power Engineering Society Transmission and Distribution Conference, 1991. With permission.

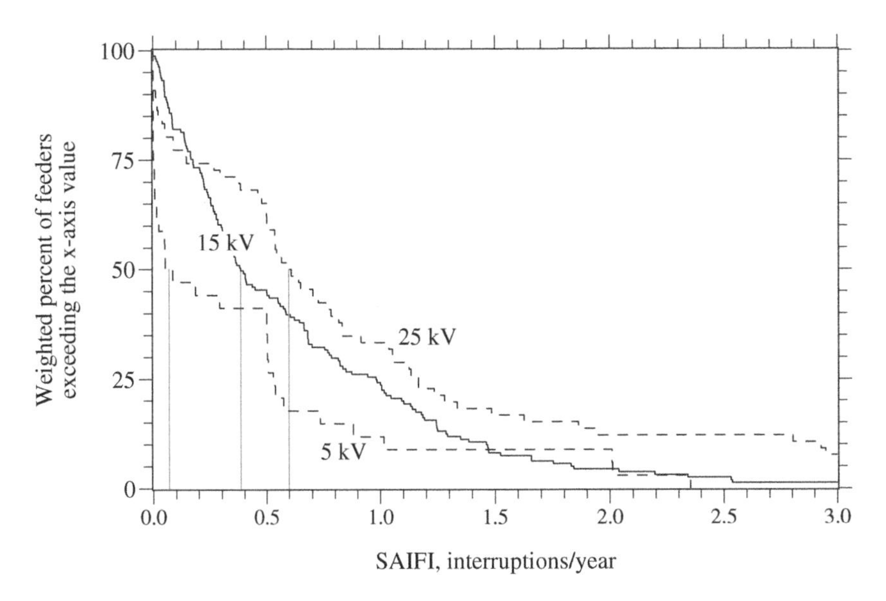

FIGURE 1.13
Effect of circuit voltage on SAIFI for one utility in the southern U.S.

TABLE 1.3
Mainline Lengths and Lateral Lengths for Optimal Reliability (Assuming a Constant Load Density)

Voltage, kV	Main Feeder Length, mi	Lateral Length, mi	Ratio of Main Feeder to Lateral Tap Length
13.8	1.51	0.95	1.59
23	1.81	1.32	1.37
34.5	2.09	1.71	1.22

Source: Burke, J. J., *Power Distribution Engineering: Fundamentals and Applications*, Marcel Dekker, New York, 1994.

Overall, the reliability trend is somewhat worsening. The main factor is probably the gradual move to higher-voltage distribution circuits and suburbanization. These trends lead to longer circuits and more exposure. Although, better record keeping (outage management systems) may be making SAIDI appear worse relative to earlier approaches because interruptions are recorded more accurately.

1.4 Modeling Radial Distribution Circuits

On purely radial circuits, the customers at the ends of the circuits unavoidably have the poorest reliability. On radial circuits, we can analyze the

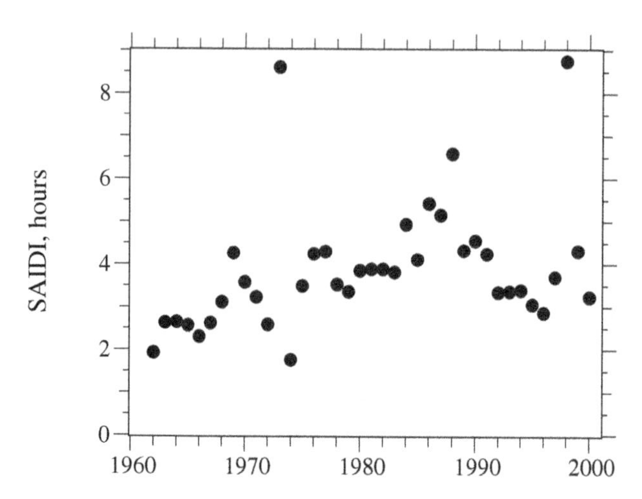

FIGURE 1.14
Yearly SAIDI for Canada. (Data from [Billinton, 1981, 2002; CEA, 2001].)

reliability using series combinations of individual elements. If any series component between the station and the customer fails, the customer loses power.

Series elements can be combined as

$$\lambda_S = \lambda_1 + \lambda_2 + \cdots + \lambda_n$$

$$U_S = U_1 + U_2 + \cdots + U_n = \lambda_1 r_1 + \lambda_2 r_2 + \cdots + \lambda_n r_n$$

$$r_S = \frac{U_S}{\lambda_S}$$

where

λ = failure rate, normally in interruptions per year

U = unavailability (total interruption time), normally in per unit, %, or h or min per year

r = average repair time per failure normally in per unit/year, %/year, or h or min

The subscript S is the total of the series combination, and the subscripts 1, 2, … n indicate the parameters of the individual elements.

The failure rate λ is analogous to SAIFI, U is analogous to SAIDI, and r is analogous to CAIDI.

We can use these basic reliability predictions to estimate reliability indices for radial circuits. Calculations quickly become complex if we try to account for sectionalizing or have circuits with parallel elements or backfeeds. Reliability analysis programs are available to model circuits with inputs similar to a load-flow program, except that switch characteristics are included as well as fault and equipment failure rates. Fault rates are the inputs most difficult to estimate accurately. These vary widely based on local conditions and construction practices.

With a given circuit configuration and SAIFI and MAIFI records for the circuit, Brown and Ochoa (1998) provide a way to back-estimate fault rates. For a given circuit configuration, the temporary and permanent fault rates are varied until the reliability prediction for a given circuit matches historical records. Once the failure rates are established, we can more accurately evaluate circuit changes such as automated switches.

1.5 Parallel Distribution Systems

To dramatically improve reliability for customers, parallel distribution supplies are needed. We see many forms of redundant distribution systems: autolooped primary distribution circuits with redundant paths, primary or secondary selective schemes with alternate supplies from two feeders, and spot or grid networks of several supply feeders with secondaries tied together. Analyzing the reliability of these interconnected systems is difficult. Several analytical techniques are available, and some are quite complicated.

With several components in series and parallel, we can find the failure rates and durations by reducing the network using the series or parallel combination of elements.

Parallel elements are combined with

$$\lambda_P = \frac{U_P}{r_p}$$

$$U_P = U_1 \times U_2 \times \cdots \times U_n = \lambda_1 \times \lambda_2 \times \cdots \times \lambda_n \times r_1 \times r_2 \times \cdots \times r_n$$

$$r_P = \frac{1}{1/r_1 + 1/r_2 + \cdots + 1/r_n}$$

for $n = 2$,

$$\lambda_P = \frac{U_P}{r_p} = \lambda_1 U_2 + \lambda_2 U_1 = \lambda_1 \lambda_2 (r_1 + r_2)$$

The subscript P is the total of the parallel combination. Note that the units must be kept the same: λ has units of 1/years, so the repair time, r, must be in units of years. Normally, this means dividing r by 8,760 if r is in hours or 525,600 if r is in minutes.

Including parallel elements is more complicated than series elements. The above equations are actually approximations that are valid only if the repair time is much less than the mean time between failure. This is generally true of distribution reliability applications (and more so for high-reliability applications).

The main problem with the equations representing the parallel combination of elements is that they are wrong for real-life electric supply reliability with multiple sources. A good illustration of this is from the data of the reliability of the utility supply found in a survey published in the *Gold Book* (IEEE Std. 493-1997). The average reliability of single-circuit supplies in the *Gold Book* has the following failure rate and repair time:

$$\lambda = 1.956 \text{ failures/year}$$

$$r = 79 \text{ min}$$

If a system were supplied with two parallel sources with the above failure rate characteristics, one would expect the following failure rates according to the ideal equations:

$$\lambda_P = (1.956)(1.956)(79+79)/525600 = 0.00115 \text{ failures/year}$$

$$r_P = 1/(1/79 + 1/79) = 39.5 \text{ min}$$

The actual surveyed reliability of circuits with multiple supplies is:

$$\lambda = 0.538 \text{ failures/year}$$

$$r = 22 \text{ min}$$

Another set of data for industrial supplies is shown in Figure 1.15 for the reliability of transmission supplies of Alberta Power. In this case, interruptions were defined as taking place for longer than 1 min. As with the *Gold Book* data, the multi-circuit Alberta Power supplies had better reliability than single-circuit supplies, but not by orders of magnitude. Single-circuit supplies had a 5-year average SAIFI of 0.9 interruptions per year (and SAIDI = 70 min of interruption/year). Multi-circuit supplies had a 5-year SAIFI of 0.42 interruptions per year (and SAIDI = 35 min of interruption/year). This information also helps show distribution engineers the number of transmission failures.

The failure rates with multiple circuits are reduced, but they are nowhere near the predicted value that is orders of magnitude lower. The reason the calculations are wrong is that the equations assume that the failures are totally independent. In reality, failures can have dependencies. The major factors are that

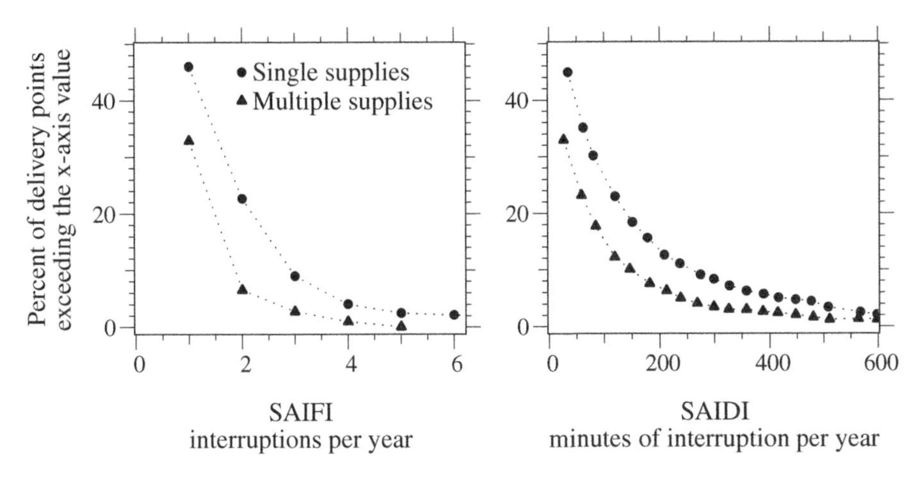

FIGURE 1.15
Comparison of single and multiple circuits for transmission supply. (Data from [Chowdhury and Koval, 1996].)

- Facilities share common space (utilities run two circuits on one structure).
- Separate supplies contain a common point upstream.
- Failures bunch together during storms.
- Maintenance must be considered.
- Hidden failures can be present.

Also, parallel supplies in many cases contain endpoint equipment that is not paralleled, including transformers, buswork, breakers, and cables.

It is possible to analytically model each of these effects. The problem is that much of the necessary input data is unknown, so many "educated" guesses are needed. For example, to analytically handle storm failures, one needs to find a storm failure rate and the duration of storms (both of these numbers are hard to come by).

It is common for multiple transmission and distribution circuits to be run on the same structures (and to be in proximity in underground facilities). If a car knocks down a pole with two circuits, both are lost. Lightning often causes multiple interruptions on structures with multiple circuits. There is also common space at the endpoint where the circuits are brought together at the customer. If there is a fire in the basement electrical room of a building with a spot network, the building will lose power.

Cable circuits are susceptible to a different type of outage bunching caused by overload failures. Cables are more sensitive to thermal overloads than are overhead lines. During high load, multiple cables can fail from overloads at the same time (and if one cable fails, the load on the others increases to make up the difference).

If a customer is supplied with two utility feeds off different feeders, the common point may be the distribution substation transformer. If the transformer or subtransmission circuit fails, both distribution feeders are lost simultaneously.

We can model common-mode failures of parallel supplies with the following equations:

$$\lambda_P = \lambda_1\lambda_2(r_1 + r_2) + \lambda_{12}$$

$$r_P = \frac{\lambda_1\lambda_2 r_1 r_2 + \lambda_{12} r_{12}}{\lambda_P}$$

where λ_{12} and r_{12} are the frequency and repair time of the common-mode failures. The effect of common-mode failures is dramatic. If they are 10% of the individual failure rates, the common-mode failures dominate and degrade the parallel-combination failure rate.

One of the main problems is that there is little data on the common-mode failure rate of utility circuits (especially distribution circuits; some data is available for transmission circuits).

Failures requiring long-duration repair times are a special case with significant impacts. If something fails and takes two months to repair or replace, the system is much more vulnerable to failures during the maintenance. Long-duration repairs also violate the approximation that the repair time is much less than the MTBF, so the normal parallel and series combination equations are in error.

Also, on systems with redundancy, human nature acts to reduce the redundancy by increasing the repair time. Repairs generally take longer because people do not feel the urgency to make the repair. If sites are without power because of a failure, there are direct, immediate consequences. With redundancy, if a failure in one component occurs, the consequences are indirect (an increase in the likelihood of failure), so repair is not as urgent. If a network primary cable fails, customers are not without power, so crews have less urgency to repair the cable.

Failures due to lightning, wind, and rain occur during storms, and the failure rate during storms is much higher than normal. Since storms are a small portion of the total time, storm interruptions are bunched together, which dramatically raises the possibility of overlapping outages. Billinton and Allan (1984) provide ways to model the effects of storm bunching, and some reliability programs model these effects.

Hidden failures do not immediately show up or show up only when a failure occurs. Some examples might include the following:

- If a customer is served with a primary selective scheme, but the transfer switch is not working, the failure remains hidden until the switch is called upon to operate.

- A five-feeder grid network was originally designed to be able to lose two feeders and still serve the load. Subsequent load growth has reduced the redundancy so that four feeders must be on-line to handle the load. If more than one feeder is lost due to a failure and/or maintenance, the load could have an interruption.

Hidden failures are difficult to track down. In a parallel distribution system, the redundancy can mask failures. Protective equipment and diagnostics equipment that can isolate or identify failures is especially helpful in reducing hidden failures.

Hidden failures are difficult to model because they are obviously hidden. Hidden failures violate the assumption that the repair time is much shorter than the MTBF. This makes it difficult to use the equations listed above for parallel and series connections to analyze hidden failures. Finally, data on hidden failures is very limited.

When designing a redundant distribution system to one or more customers, consider these strategies that help reduce the possibility of overlapping failures:

- *Common space* — Limit sharing of physical space as much as possible. For services with multiple sources — primary or secondary selective schemes — try to use circuits that do not share the same poles or even right of ways. Use circuits that originate out of different distribution substations.
- *Storms* — Using underground equipment helps reduce the possibility of overlapping storm outages (although cables have their own bunched failures from overloads).
- *Maintenance* — Coordinate maintenance as much as possible to limit the loss of redundancy during maintenance. Try to avoid maintenance during stormy weather including heat waves for cables.
- *Testing* — Test switches, protective devices, and other equipment where hidden failures may lurk.
- *Loadings* — Review loadings periodically to ensure that overloads are not reducing designed redundancy.

1.6 Improving Reliability

We have many different methods of reducing long-duration interruptions including

- *Reduce faults* — tree trimming, tree wire, animal guards, arresters, circuit patrols

- *Find and repair faults faster* — faulted circuit indicators, outage management system, crew staffing, better cable fault finding
- *Limit the number of customers interrupted* — more fuses, reclosers, sectionalizers
- *Only interrupt customers for permanent faults* — reclosers instead of fuses, fuse saving schemes

Whether we are trying to improve the reliability on one particular circuit or trying to raise the reliability system wide, the main steps are

1. Identify possible projects.
2. Estimate the cost of each configuration or option.
3. Estimate the improvement in reliability with each option.
4. Rank the projects based on a cost-benefit ratio.

Prediction of costs is generally straightforward; predicting improvement is not. Some projects are difficult to attach a number to.

An important step in improving reliability is defining what measure to optimize: is it SAIFI, SAIDI, some combination, or something else entirely? The ranked projects change with the goal. Surveys have shown that frequency of interruptions is most important to customers (until you get to very long interruptions). Regulators tend to favor duration indicators since they are more of an indicator of utility responsiveness, and excessive cost cutting might first appear as a longer response time to interruptions.

Detailed analysis and ranking of projects can be done on a large scale. Brown et. al. (2001) provide an interesting example of applying reliability modeling to Commonwealth Edison's entire distribution system in Illinois to rank configuration improvements. Normally, large-scale projects require simplification (and often a good bit of guesswork).

Adding reclosers, putting in more fusing points, automating switches — these configuration changes are predictable. Many computer programs will quantify these improvements. Projects aimed at reducing the rates of faults, such as trimming more trees, adding more arresters, installing squirrel guards, are difficult to quantify. Improving fault-finding and repair are also more difficult to quantify. A sensitivity analysis helps when deciding on these projects. In the simplest form, rather than using one performance number, use a low, a best guess, and a high estimate. Pinpointing fault causes also helps frame how much benefit these targeted solutions can have (if there are few lightning-caused faults, additional arresters will provide little benefit).

1.6.1 Identify and Target Fault Causes

Tracking and targeting fault types helps identify where to focus improvements. If animals are not causing faults, we do not need animal guards.

Many utilities tag interruptions with identifying codes. The system-wide database of fault identifications is a treasure of information that we can use to help improve future reliability.

Different fault causes affect different reliability indices. Figure 1.16 shows the impact of several interruption causes on different reliability parameters for Canadian utilities. Relative impacts vary widely; for example, trees had a high repair time but impacted fewer customers.

Tracking this type of data for a utility operating region helps identify the most common problems for that service area. These numbers change by region depending on weather, construction practices, load densities, and other factors.

1.6.2 Identify and Target Circuits

Do not treat all circuits the same. The most important sections are usually not the locations with the most faults per mile. The number of customers on a circuit and the type of customers on a circuit are important considerations. For example, a suburban circuit with many high-tech commercial customers should warrant different treatment than a rural circuit with fewer, mostly residential and agricultural customers. How this is weighted depends on the utility's philosophy.

On radial distribution circuits, the three-phase mainline is critical. Sustained interruptions on the mains locks out all customers on the circuit until crews repair the damage. Feeders with extra-long mainlines have more interruptions. To reduce the impact of mainline exposure, trim trees more often on the mains and patrol the mains more often. Mainline sectionalizing switches help by allowing quick restoration of customers upstream of the fault; automated switches are even better. Another improvement is using normally open tie switches to other feeders, enabling crews to move load to other feeders during sectionalizing.

Lateral taps are another target. We can rank these by historical performance, taking into account their length and number of customers. Some longer laterals are good candidates for single-phase reclosers instead of fuses.

1.6.3 Switching and Protection Equipment

Fuses, sectionalizing switches, reclosers, sectionalizers — more, more, more — the more we have, the more we isolate faults to smaller chunks of circuitry, the fewer customers we interrupt.

Taps are almost universally fused, primarily for reliability. Fuses make cheap fault finders. Planners should also try to design to have tap exposure, not too much and not too little. We want to have a high percentage of a circuit's exposure on fused taps, so when permanent faults occur on those sections, only a small number of customers are interrupted.

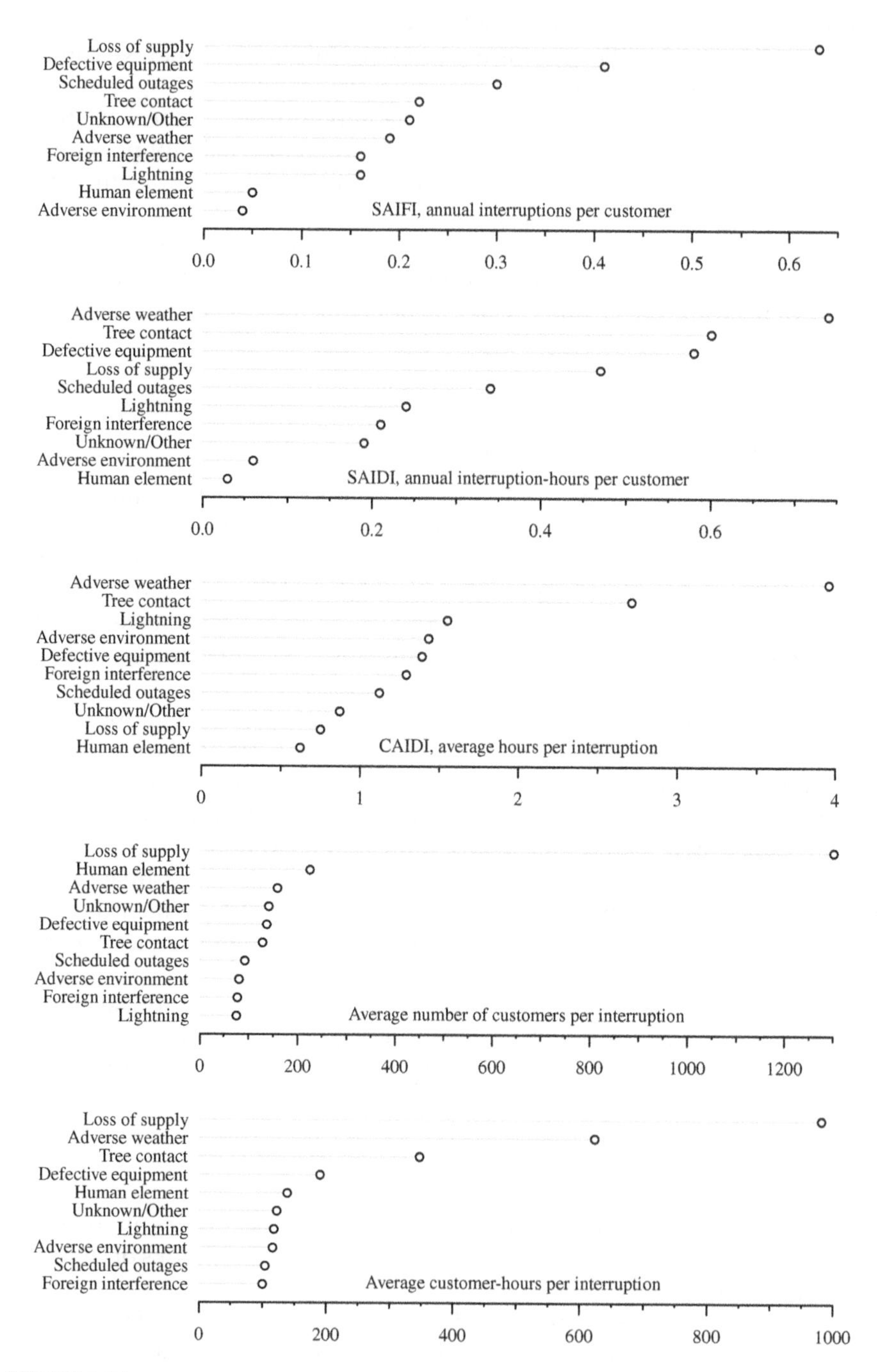

FIGURE 1.16
Root-cause contributors to different reliability parameters. (Data from [CEA, 2001].)

TABLE 1.4

Example Reliability Improvement Calculations

	SAIFI	MAIFI
Base case, fuse saving	5(0.3)+0.5(0.3) = 1.65	5(0.5)+10(0.9) = 12
Base case, fuse blowing	5(0.3)+0.5(0.9) = 1.95	5(0.6) = 3
One recloser, fuse blowing	(1.95+1.95/2)/2 = 1.46	(3+3/2)/2 = 2.25
Three-recloser auto-loop, fuse blowing	1.95/2 = 0.98	(3+3/2)/2 = 2.25
Five-recloser auto-loop, fuse blowing	1.95/3 = 0.65	(3+2+1)/3 = 2

Note: 5-mi mains, 15 total mi of exposure, 0.3 permanent faults/mi/year, 0.6 temporary faults/mi/year, fused laterals average 0.5 mi, customers are evenly distributed along the mainline and taps.

If taps become too long, use reclosers instead of fuses. Especially for circuits that fan out into two or three main sections (really like having two or three mainlines), reclosers on each of the main sections help improve reliability.

How circuits are protected and coordinated impacts reliability. Fuse saving, where the station breaker trips before tap fuses to try to clear temporary faults, helps long-duration interruptions most (but causes more momentary interruptions). Fuse blowing causes more long-duration interruptions because the fuse always blows, even for temporary faults.

Mainline reclosers also help improve reliability. Table 1.4 compares different scenarios for a common feeder with the following assumptions: 5-mi mains, 15 total mi of exposure, 0.3 permanent faults/mi/year, 0.6 temporary faults/mi/year, fused laterals average 0.5 mi, and customers are evenly distributed along the main line and taps. Mainline faults contribute most to SAIFI (1.5 interruptions per year). If the system were not fused at all, it would have 4.5 interruptions per year, pointing out the great benefit of the fuses. Branch-line faults only contribute an average of 0.15 interruptions with fuse saving (assuming fuse saving works right). This is an average; some customers on long taps have many more interruptions due to branch-line faults (these are good candidates for reclosers instead of fuses). On a purely radial system, reclosers do not help customers at the end of the line. An auto-loop scheme helps the customers at the ends of the line and significantly improves the feeder reliability indices.

This example only includes distribution primary interruptions. Supply-side interruptions and secondary interruptions should also be added as appropriate.

Sectionalizing switches can significantly improve SAIDI and CAIDI (but not SAIFI, unless the switches are automated). Such switches enable crews to easily reenergize significant numbers of customers well before they fix the actual damage. As Brown's analysis (2002) shows, the biggest gains are with the first few switches. Both SAIDI and CAIDI for mainline faults reduce in proportion to the difference between the mean time to repair (t_{repair}) and the mean time to switch (t_{switch}). With evenly spaced switches and customers on a radial circuit with no backfeeds, both CAIDI and SAIDI reduce with increasing numbers of sectionalizing switches as

$$R = \frac{\sum_{i=1}^{n} i}{(n+1)^2}(t_{repair} - t_{switch})$$

where

R = reduction in CAIDI and SAIDI
n = number of sectionalizing switches on the mainline
t_{repair} = time to repair the damage
t_{switch} = time to operate the sectionalizing switch

On a radial circuit, the improvement is not distributed equally; switches do not help customers at the end of the line at all. If a circuit has two evenly spaced sectionalizing circuits, the customers on the last part of the circuit see no improvement, the middle third see improvement for faults on the last third of the circuit, and the third at the front see improvement for faults on the last two-thirds of the circuit.

If the circuit has a backfeed from another circuit, the improvement is equal for all customers, so the overall circuit CAIDI improves. For evenly spaced switches and customers, the reduction is

$$R = \frac{n}{n+1}(t_{repair} - t_{switch})$$

Figure 1.17 shows how much sectionalizing switches can reduce CAIDI and SAIDI for the portion due to faults on the main line. The first switches provide the biggest bang for the buck. Beyond five switches, improvement is marginal. For application on real feeders (where the loads are not evenly placed), for biggest gains on radial circuits, place switches just downstream of large blocks of customers. If a circuit has a branch line with many customers, a mainline sectionalizing switch just downstream of the tap point allows crews to sectionalize that big block of customers for faults downstream of the switch.

The mean time to switch (t_{switch}) includes the time for the crew to get to the circuit, the time they need to find the fault, the time to find and open the appropriate sectionalizing switch, and finally the time to close the tripped breaker or recloser. Typically, t_{switch} is about 1 h. The mean time to repair (t_{repair}) is the time to travel to the circuit, find the fault, repair the damage, and close in the appropriate switching device to reconnect customers. The repair time varies widely — 4 h is a good estimate; actual repairs regularly range from 2 to 8 h.

Crews should decide whether to sectionalize based on local conditions. What is damaged? How long will it take to fix? How many customers would

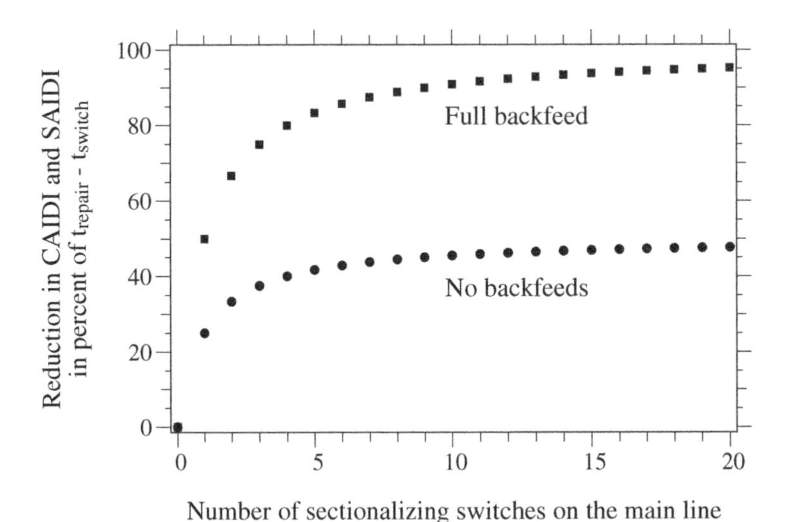

FIGURE 1.17
Reducing interruption durations with normally closed sectionalizing switches on the mainline.

sectionalizing bring back? Where are the sectionalizing switches? How long will it take to sectionalize the circuit? Sometimes, crews can start sectionalizing before the fault is found. If crews patrol a circuit section and do not see damage, they can open a sectionalizing switch and reclose the station breaker before they continue looking for the fault downstream.

1.6.4 Automation

Automation provides options for improving the reliability of the distribution supply. An auto-loop automated distribution configuration is a popular way to improve reliability on a normally radial circuit. These systems automatically reconfigure a distribution system: we do not need outside intervention or communications. In the three-recloser loop example in Figure 1.18, a normal sequence of operations for a fault upstream of recloser R1 is: (1) breaker B1 senses the fault and goes through its normal reclosing cycle and locks open; (2) recloser R1 senses loss of voltage and opens; and (3) recloser R2, the tie recloser, senses loss of voltage on the feeder and closes in. Since R2 can be switching into a fault, normally it is set for one shot; if the fault is there, it trips and stays open.

We can add more reclosers to divide the loop into more sections, but coordination of all of the reclosers is harder. Consider a five-recloser loop (Figure 1.19). Each feeder has two normally closed reclosers, and there is a normally open tie-point recloser. If feeder #1 is faulted close to the substation, breaker B1 locks out, recloser R1 opens, and the tie recloser closes. Now, we have a long radial circuit with the station breaker in series with four reclosers — that is a lot to try to coordinate. To ease the coordination, some reclosers can lower their tripping characteristics when operating in reverse mode. So,

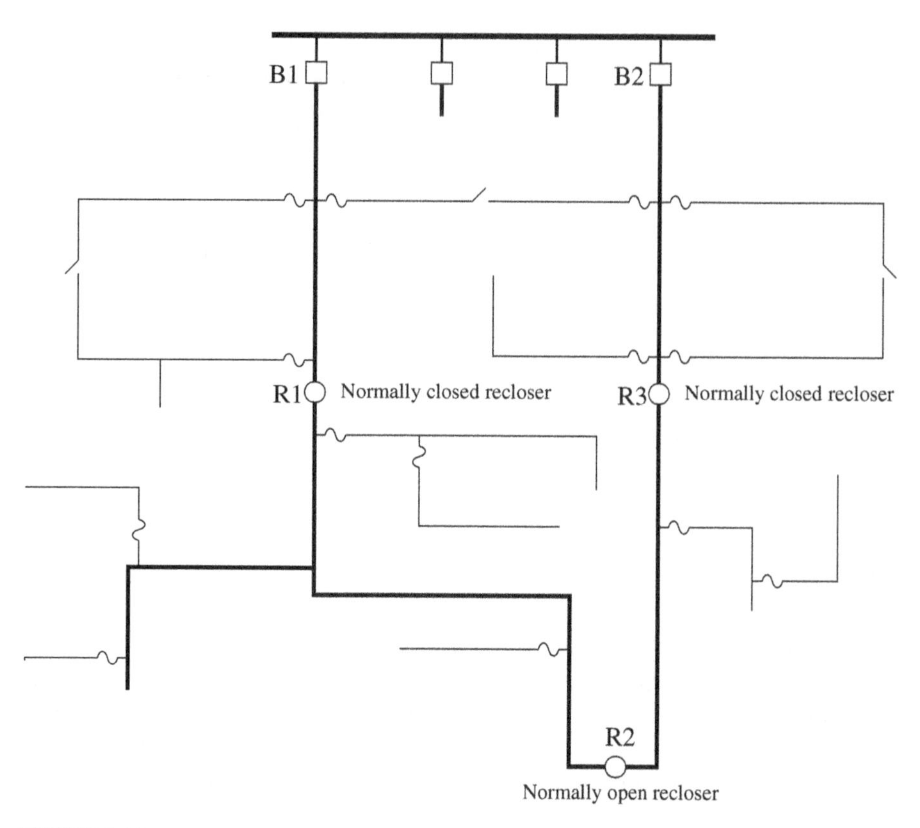

FIGURE 1.18
Example of an automated distribution feeder.

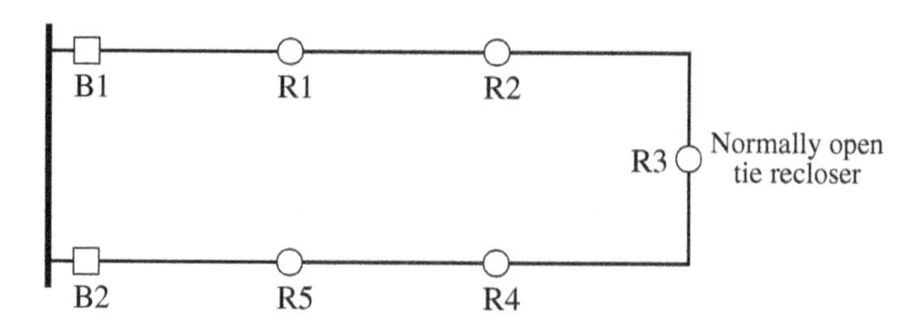

FIGURE 1.19
Five-recloser automated loop.

in this example, recloser R2 would drop its pickup setting. R2 sees much lower fault currents than it usually does, and we want it to trip before recloser R3 or R4.

For a fault between B1 and R1, the five-recloser loop responds similarly to a three-recloser loop: (1) breaker B1 locks out, (2) R1 opens on loss of voltage, (3) recloser R2 drops its trip setting, and (4) R3 senses loss of voltage on feeder 1 and closes in. For a fault between R1 and R2, the sequence is more complicated: (1) recloser R1 locks out, (2) recloser R2 drops its trip setting and goes to one shot until lockout, (3) R3 senses loss of voltage on feeder 1 and closes in (and closes in on the fault), and (4) R2 trips in one shot due to its lower setting. In a variation of this scheme, utilities use sectionalizers instead of reclosers at positions R2 and R4. Sectionalizers are easier to coordinate with several devices in series.

Remotely controlled switches are another option for automating a distribution circuit. The preferred communication is radio. Remotely controlled switches are more flexible than auto-loop schemes because it is easier to apply more tie points and we do not have to worry about coordinating protective equipment. Most commonly, operators decide how to reconfigure a circuit.

Even if a circuit is automated, doing another step of sectionalizing within the isolated section can squeeze out better reliability. Brown and Hanson (2001) show that manually sectionalizing after automated switches have operated can reduce SAIDI by several percent. As with other sectionalizing, crews should decide on a case-by-case basis whether to sectionalize.

Auto-loops will not necessarily help with momentary interruptions. Automation turns long-duration interruptions into momentary interruptions. To help with momentary interruptions, consider the following enhancements to automation schemes:

- *Line reclosers* — As part of an automated loop, line reclosers significantly improve momentaries; automated switches do not. Using single-phase reclosers helps interrupt fewer customers.
- *Tap reclosers* — Use reclosers on long lateral taps. Consider single-phase reclosers on three-phase taps. These will interrupt fewer customers.

1.6.5 Maintenance and Inspections

For many utilities, the best maintenance is trimming trees and then trimming some more trees; tree trimming is by far the largest maintenance expense for these utilities. Beyond tree trimming, distribution circuit maintenance practices vary widely. Distribution transformers, capacitors, insulators, wires, cables — most distribution equipment — do not need maintenance. Oil-filled switches, reclosers, and regulators need only occasional maintenance.

Most maintenance involves identifying old and failing equipment and targeting it for replacement. Equipment deteriorates over time. Several utilities have increasingly older infrastructure. Equipment fails at varying rates over its lifetime. Typically, it is a "bathtub curve": high failure rates initially during the break-in period (mostly due to manufacturing defects), a period of "normal" failure rates that increases over the equipment's lifetime. Some equipment sees more acceleration in failure rates than others. Data for distribution equipment is difficult to find. Early plastic cables — high-molecular weight polyethylene and cross-linked polyethylene — had dramatically increasing failure rates. Duckett and McDonough (1990) found dramatically increasing failure rates with age on Carolina Power & Light's 14.4-kV, 125-kV BIL transformers based on failures recorded from 1984 through 1988. The failure rate shot up when units reached 15 to 20 years old. An earlier study of CP&L's 7.2-kV, 95-kV BIL transformers did not show an increasing failure rate with age, staying at about 0.2 to 0.4% annually (Albrecht and Campbell, 1972). Aged transformers are more susceptible to failure, but we cannot justify replacement based on age; cables are the only equipment that utilities routinely replace solely based on age.

Storms trigger much "maintenance" — storms knock lines and equipment down, and crews put them back up (this is really restoration).

From birth to death, tracking equipment quality and failures helps improve equipment reliability. On most overhead circuits, most failures are external causes, not equipment failures (usually about 10 to 20% are equipment failures). Still, tracking equipment failures and targeting "bad apples" helps improve reliability. Many utilities do not track equipment failures at all. But, some utilities have implemented programs for tracking equipment and their failures. Failures occur in clusters: particular manufacturers, particular models, particular manufacturing years. Whether it is a certain type of connector or a brand of standoff insulator, some equipment has much higher than expected failure rates.

Proper application of equipment also helps, especially not overloading equipment excessively and applying good surge protection.

On underground circuits, equipment failures cause most interruptions. Tracking cable failures (usually by year of installation and type of insulation) and accessory failures and then replacing poor performers helps improve reliability. Monitoring loadings helps identify circuits that may fail thermally.

Quality acceptance testing of new equipment, especially cables, can identify poor equipment before it enters the field. For cables, tests can include microscopic evaluation of slices of cables to identify voids and impurities in samples. A high-pot test can also identify bad batches of cable.

On underground circuits, since workmanship plays a key role in quality of splices, tracking can also help. If a splice fails 6 months after it is installed and if we know who did the splice and who made the splice, we can work to correct the problem, whether it was due to workmanship or poor manufacturing quality.

Utilities use a variety of inspection programs to improve reliability. Of North American utilities surveyed (CEA 290 D 975, 1995), slightly more than half have regular inspection programs, and fewer than 5% have no inspections. Efforts varied widely: 27% spent less than 2% of operations and maintenance budgets on inspections, while 16% of utilities spent 10 to 30% of O&M on inspections.

Some distribution line inspection techniques used are

- *Visual inspections* — Most often, crews find gross problems, especially with drive-bys: severely degraded poles, broken conductor strands, and broken insulators. Some utilities do regular visual inspections; but more commonly, utilities have crews inspect circuits during other activities or have targeted inspections based on circuit performance. The most effective inspections are those geared towards finding fault sources — these may be subtle; crews need to be trained to identify them (see chapter 4).

- *Infrared thermography* — Roughly 40% of utilities surveyed use infrared inspections for overhead and underground circuits. Normally, crews watch a 20°C rise and initiate repair for more than a 30°C rise. Infrared scanning primarily identifies poor connectors. Some utilities surveyed rejected infrared monitoring and did not find it cost effective. Other utilities found significant benefit.
- *Wood pole tests* — Visual inspections are most common for identifying weak poles. A few utilities use more accurate measures to identify the mechanical strength left in poles. A hammer test, whacking the pole with a sledge, is slightly more sophisticated; a rotted pole sounds different when compared to a solid pole. Sonic testing machines are available that determine density and detect voids.
- *Operation counts* — Most utilities periodically read recloser operation and regulator tap changer counters to identify when they need maintenance.
- *Oil tests* — A few utilities perform oil tests on distribution transformers, reclosers, and/or regulators. While these tests can detect deterioration through the presence of water or dissolved gasses, the expense is difficult to justify for most distribution equipment.

Substation inspections and maintenance are more universally accepted. Most utilities track operation counts or station breakers and regulators, and most also sample and test station transformer oil periodically.

1.6.6 Restoration

Restoration affects SAIDI and CAIDI. Repair times vary considerably as shown in the example in Figure 1.20. Response time degrades quickly during storms as all crew resources are locked up. Even if "major events" are excluded, the responsiveness during bad weather still greatly influences restoration time.

The main way to improve restoration time is to sectionalize the circuit to bring as many customers back in as quickly as possible. Other methods that help reduce the repair time include the following:

- *Prepare* — Use weather information including lightning detection networks to track storms. Call out crews before the interruptions hit. Coordinate crews to distribute them as efficiently as possible.

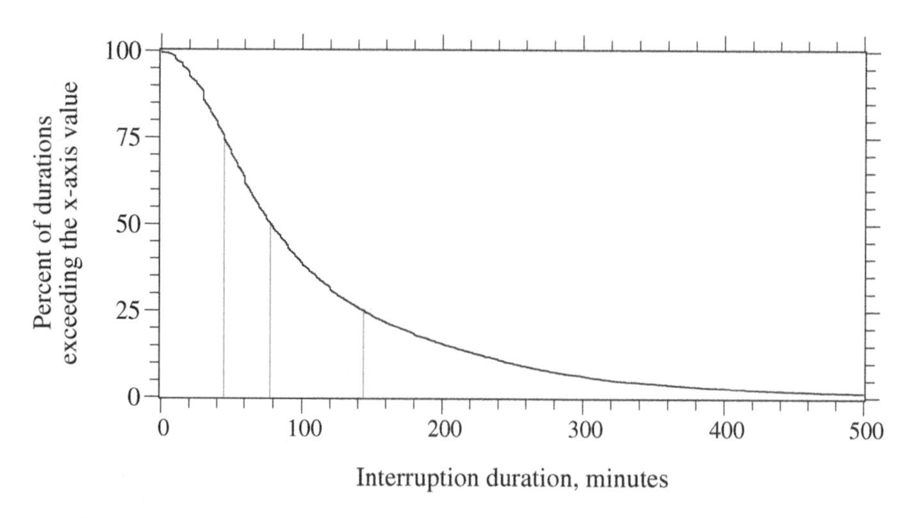

FIGURE 1.20
Distribution of interruption durations at one utility. (Data from [IEEE Working Group on System Design, 2001].)

- *Train* — Storm response training and other crew training help improve responsiveness.
- *Locate* — Use faulted circuit indicators and better cable locating equipment; have better system maps available to crews patrolling circuits. Use more fuses — a fuse is a cheap fault locator; with a smaller area downstream of a fuse, less length needs patrolling. Better communication between the call center and crews helps send crews to the right location.
- *Prioritize* — During storms, prioritize efforts based on those that get the most customers back in service quickly. Many of the first efforts will be sectionalizing; next will be efforts to target the repairs affecting most customers (faults on the distribution system mainline). Downed secondary and other failures affecting small numbers of customers have to wait. When prioritizing, safety implications should override reliability concerns; make sure downed wire cases are deenergized before other repairs.
- *Target* — Apply maintenance to address the faults that require long repair times. Tree faults have long repair times, so tree trimming reduces the repair time.

An outage management system helps with restoration and gives utilities information to help improve performance. But, be aware that implementing an outage management system will normally make reliability indices worse — just the indices. The actual effect on customers is not worse; in fact it should improve as utilities use the outage management system to improve responsiveness. Unfortunately, better record keeping translates into higher reliability indices. Several utilities have reported that SAIFI and SAIDI increase between 20 and 50% after implementing an outage management system. Outage management systems do help improve reliability and efficiency. Responsiveness improves as outage information is relayed more directly to crews. Outage management systems also calculate the reliability indices for utilities and can generate reports that utilities can use to target certain circuits for inspections or tree trimming. In addition to reliability, customer satisfaction improves, as call centers (either automated or people operated) are able to give customers better information on restoration times.

Knowing when most storms tend to occur and when most interruptions occur helps for scheduling crews. Typically, summer months are the busiest. Of course, each area has somewhat different patterns. Figure 1.21 shows SAIDI data from four U.S. utilities. Both a median and an average are shown — the median represents a typical day; the average counts towards the yearly index. One or two severe storm days can appreciably raise the average for the given month. Some utilities are hit much more by storms (those with high ratios of average to median).

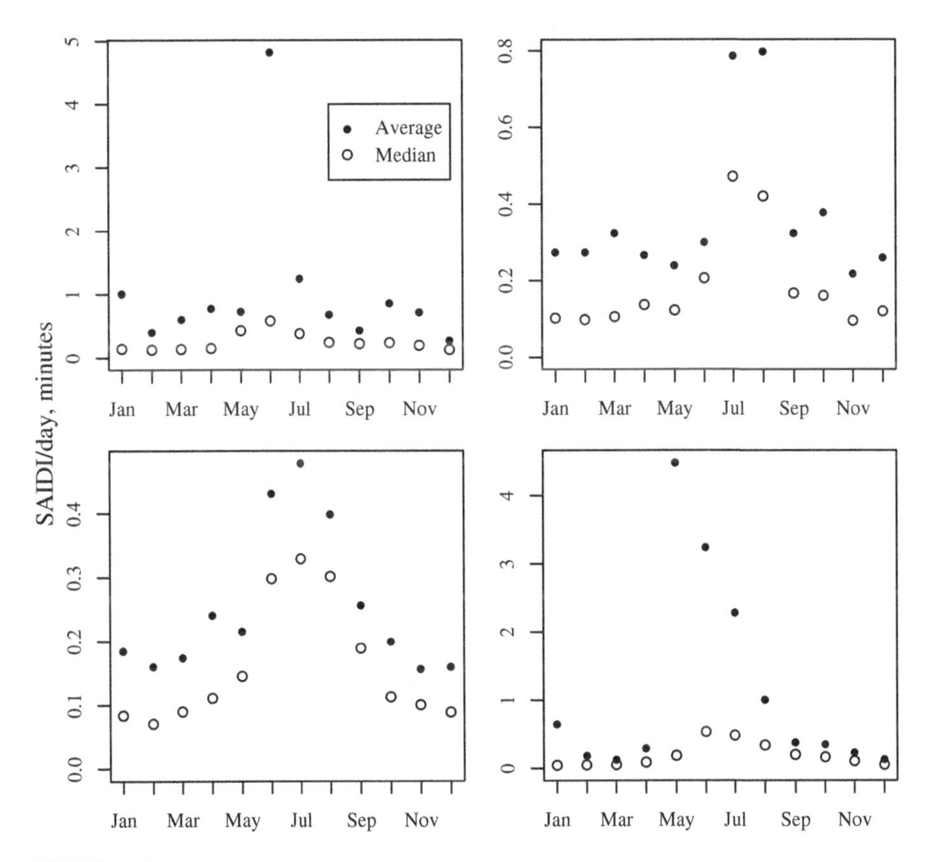

FIGURE 1.21
SAIDI per day by month of the year for four utilities.

Safety: Remember safety. Always. Reliability is important, but not worth dying for. Do not push repairs so quickly that crews take shortcuts that might create dangerous situations. Tired crews and rushed crews make more mistakes. Do not work during active lightning storms or other dangerous conditions. Make sure that the right people are doing the job; make sure they use the right tools, take enough breaks, and follow normal safety precautions.

1.6.7 Fault Reduction

An obvious approach to reliability improvement is to reduce the number of faults. In addition to long-duration interruptions, this strategy reduces the number of voltage sags and momentary interruptions and makes the system safer for workers and the public.

On-site investigations of specific faults can help reduce subsequent faults. Faults tend to repeat at the same locations and follow patterns. For example, one particular type and brand of connector may have a high failure rate. If these are identified, replacement strategies can be implemented. Another example is animal faults — one particular pole that happens to be a good travel path for squirrels may have a transformer with no animal guards. The same location may have repeated outages. These may be difficult to find at first, but crews can be trained to spot pole structures where faults might be likely.

1.7 Interruption Costs

Damaged equipment, overtime pay, lost sales, damage claims from customers — interruptions cost utilities money, plenty of money. An EPRI survey found that an average of 10% of annual distribution costs are for service restoration with ranges at different utilities from 7.6 to 14.8% (EPRI TR-109178, 1998). Restoration averaged $14 per customer and $0.20 per customer minute of interruption. The $14 is per customer, not per customer interrupted, but it is close for a typical SAIFI of 1 to 1.5; assuming SAIFI equals 1.4, the average restoration cost is $10 per customer interruption. Table 1.5 shows the restoration costs scaled by several factors. Not surprisingly, most of the cost is the actual construction to fix the problem as shown in Figure 1.22. Also, labor is the biggest portion of restoration costs, more than 70% in the survey. Note that the costs reported in the survey are costs directly associated with the restoration; lost kWh sales and damage claims are not included.

Costs escalate for major storms that severely damage distribution infrastructure. Table 1.6 shows the Duke Power Company's costs for several major storms. Many of these storms had much higher than normal costs per customer interrupted (as well as very high absolute costs).

TABLE 1.5

Surveyed Utility Restoration Costs in U.S. Dollars

	Average	Range
Per customer	14	12–17
Per customer minute of interruption	0.2	0.16–0.27
Per mile of overhead circuit	1000	300–1850
Per mile of underground circuit	3100	1700–5500

Source: EPRI TR-109178, *Distribution Cost Structure — Methodology and Generic Data,* Electric Power Research Institute, Palo Alto, CA, 1998.

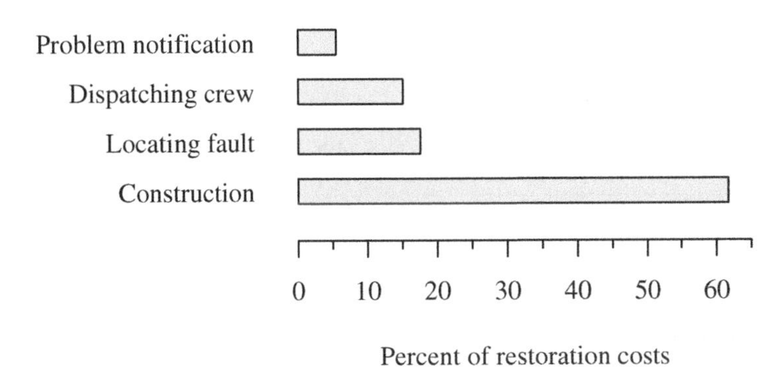

FIGURE 1.22
Breakdown of utility restoration costs. (Data from [EPRI TR-109178, 1998].)

TABLE 1.6

Restoration Costs During a Typical Year and During Major Storms for the Duke Power Company

Date	Storm Type	Customers Interrupted	Cost, $k	Cost Per Customer Interrupted, $
May-89	Tornadoes	228,341	15,190	67
Sep-89	Hurricane Hugo	568,445	64,671	114
Mar-93	Wind, Ice, and Snow	146,436	9,176	63
Oct-95	Hurricane Opal	116,271	1,655	14
Jan-96	Western NC Snow	88,076	873	10
Feb-96	Ice Storm	660,000	22,906	35
Sep-96	Hurricane Fran	409,935	17,472	43

Source: Keener, R. N., "The Estimated Impact of Weather on Daily Electric Utility Operations," Social and Economic Impacts of Weather, Proceedings of a workshop at the University Corporation for Atmospheric Research, Boulder, CO, 1997. Available at http://sciencepolicy.colorado.edu/socasp/weather1/keener.html.

TABLE 1.7

Survey of Interruption Costs to 299 Large Commercial and Industrial Customers

	4-h Interruption, No Notice	1-h Interruption, No Notice	1-h Interruption with Notice	Momentary Interruption	Voltage Sag
Production time lost, hours	6.67	2.96	2.26	0.70	0.36
Percent of work stopped	91%	91%	91%	57%	37%
Average total costs	$74,835	$39,459	$22,973	$11,027	$7,694
Costs per monthly kWh	0.2981	0.0182	0.0438	0.0506	0.0492

Source: Sullivan, M. J., Vardell, T., and Johnson, M., "Power Interruption Costs to Industrial and Commercial Consumers of Electricity," *IEEE Transactions on Industry Applications*, vol. 33, no. 6, pp. 1448–58, November/December 1997.

TABLE 1.8

Survey of Interruption Costs to Puget Sound Energy Customers

	12-h Interruption	4-h Interruption	1-h Interruption	Momentary Interruption
Commercial and industrial	$5144	$2300	$1008	$109
Residential	$25.95	$12.73	$8.32	$3.64

Source: Sullivan, M. and Sheehan, M., "Observed Changes in Residential and Commercial Customer Interruption Costs in the Pacific Northwest Between 1989 and 1999," IEEE Power Engineering Society Summer Meeting, 2000.

Some utilities also consider the costs to customers when planning for reliability. Costs of interruptions for customers vary widely, depending on the type of customer, the size of customer, the duration of the interruption, and the time of day and day of the week. Costs are highest for large commercial and industrial customers — Table 1.7 shows averages of customer costs for large commercial and industrial customers for various interruptions and short-duration events. Costs rise for longer-duration interruptions. Table 1.8 shows surveyed interruption costs for one utility's customers. We have to be careful of surveyed results of reliability surveys; utility customers often will not actually pay for solutions to eliminate the interruptions, even if the solution has a very short payback assuming their claimed costs of losses.

References

Albrecht, P. F. and Campbell, H. E., "Reliability Analysis of Distribution Equipment Failure Data," EEI T&D Committee, New Orleans, LA, January 20, 1972. As cited by Duckett and McDonough (1990).

Billinton, R., "Comprehensive Indices for Assessing Distribution System Reliability," IEEE International Electrical, Electronics Conference and Exposition, 1981.

Billinton, R., 2002. Personal communication.

Billinton, R. and Allan, R. N., *Reliability Evaluation of Power Systems*, Pitman Advanced Publishing Program, 1984.

Brown, R. E., *Electric Power Distribution Reliability*, Marcel Dekker, New York, 2002.

Brown, R. E. and Burke, J. J., "Managing the Risk of Performance Based Rates," *IEEE Transactions on Power Systems*, vol. 15, no. 2, pp. 893–8, May 2000.

Brown, R. E., Gupta, S., Christie, R. D., Venkata, S. S., and Fletcher, R., "Distribution System Reliability Assessment: Momentary Interruptions and Storms," *IEEE Transactions on Power Delivery*, vol. 12, no. 4, pp. 1569–75, October 1997.

Brown, R. E. and Hanson, A. P., "Impact of Two-Stage Service Restoration on Distribution Reliability," *IEEE Transactions on Power Systems*, vol. 16, no. 4, pp. 624–9, November 2001.

Brown, R. E., Hanson, A. P., Willis, H. L., Luedtke, F. A., and Born, M. F., "Assessing the Reliability of Distribution Systems," *IEEE Computer Applications in Power*, vol. 14, no. 1, pp. 44–49, 2001.

Brown, R. E. and Ochoa, J. R., "Distribution System Reliability: Default Data and Model Validation," *IEEE Transactions on Power Systems*, vol. 13, no. 2, pp. 704–9, May 1998.

Burke, J. J., *Power Distribution Engineering: Fundamentals and Applications*, Marcel Dekker, New York, 1994.

CEA 160 D 597, *Effect of Lightning on the Operating Reliability of Distribution Systems*, Canadian Electrical Association, Montreal, Quebec, 1998.

CEA 290 D 975, *Assessing the Effectiveness of Existing Distribution Monitoring Techniques*, Canadian Electrical Association, 1995.

CEA, *CEA 2000 Annual Service Continuity Report on Distribution System Performance in Electric Utilities*, Canadian Electrical Association, 2001.

Chowdhury, A. A. and Koval, D. O., "Delivery Point Reliability Measurement," *IEEE Transactions on Industry Applications*, vol. 32, no. 6, pp. 1440–8, November/December 1996.

Christie, R. D., "Statistical methods of classifying major event days in distribution systems," IEEE Power Engineering Society Summer Meeting, 2002.

Coulter, R. T., "2001 Electricity Distribution Price Review Reliability Service Standards," Prepared for the Office of the Regulator-General, Victoria, Australia and Service Standards Working Group, 1999.

Duckett, D. A. and McDonough, C. M., "A guide for transformer replacement based on reliability and economics," Rural Electric Power Conference, 1990. Papers Presented at the 34th Annual Conference, General Electric Co., Hickory, NC, 1990.

EEI, "EEI Reliability Survey," Minutes of the 8th Meeting of the Distribution Committee, March 28–31, 1999.

EPRI TR-109178, *Distribution Cost Structure — Methodology and Generic Data*, Electric Power Research Institute, Palo Alto, CA, 1998.

IEEE Std. 493-1997, *IEEE Recommended Practice for the Design of Reliable Industrial and Commercial Power Systems (Gold Book)*.

IEEE Std. 1366-2000, *IEEE Guide for Electric Power Distribution Reliability Indices*.

IEEE Working Group on System Design, http://grouper.ieee.org/groups/td/dist/sd/utility2.xls, 2001.

Indianapolis Power & Light, "Comments of Indianapolis Power & Light Company to Proposed Discussion Topic, Session 7, Service Quality Issues," submission to the Indiana Regulatory Commission, 2000.

Keener, R. N., "The Estimated Impact of Weather on Daily Electric Utility Operations," Social and Economic Impacts of Weather, Proceedings of a workshop at the University Corporation for Atmospheric Research, Boulder, CO, 1997. Available at http://sciencepolicy.colorado.edu/socasp/weather1/keener.html.

MacGorman, D. R., Maier, M. W., and Rust, W. D., "Lightning Strike Density for the Contiguous United States from Thunderstorm Duration Records." Report to the U.S. Nuclear Regulatory Commission, # NUREG/CR-3759, 1984.

PA Consulting, "Evaluating Utility Operations and Customer Service," FMEA-FMPA Annual Conference, Boca Raton, FL, July 23–26, 2001.

Settembrini, R. C., Fisher, J. R., and Hudak, N. E., "Reliability and Quality Comparisons of Electric Power Distribution Systems," IEEE Power Engineering Society Transmission and Distribution Conference, 1991.

Short, T. A., "Reliability Indices," T&D World Expo, Indianapolis, IN, 2002.

Sullivan, M. and Sheehan, M., "Observed Changes in Residential and Commercial Customer Interruption Costs in the Pacific Northwest Between 1989 and 1999," IEEE Power Engineering Society Summer Meeting, 2000.

Sullivan, M. J., Vardell, T., and Johnson, M., "Power Interruption Costs to Industrial and Commercial Consumers of Electricity," *IEEE Transactions on Industry Applications*, vol. 33, no. 6, pp. 1448-58, November/December 1997.

Selected responses to: Close In Or Patrol?

Our company policy is to try to close and then patrol.

A line fuse is blown. Primary is on the ground. Children are around it. And you want to try it first! You better patrol the line!

I think you should patrol first. We once killed a farmer's cow because primary was down on his fence. Lucky it was just a cow.

Our Company Policy is to Close in on the line on arrival, if the phone center has not recieved any wire down calls. I'm not sure this is a good policy, because Mr and Mrs Customer don't know a powerline from a washline......I personally would ride the line and then when I was convinced that I have made a good call, close in.................Be CAREFUL............

www.powerlineman.com

2

Voltage Sags and Momentary Interruptions

The three most significant power quality concerns for most customers are

- Voltage sags
- Momentary interruptions
- Sustained interruptions

Different customers are affected differently. Most residential customers are affected by sustained interruptions and momentary interruptions. For commercial and industrial customers, sags and momentaries are the most common problems. Each circuit is different, and each customer responds differently to power quality disturbances. These three power quality problems are caused by faults on the utility power system, with most of them on the distribution system. Faults can never be completely eliminated, but we have several ways to minimize the impact on customers.

Of course, several other types of power quality (PQ) problems can occur, but these three are the most common; sags and momentary interruptions are addressed in this chapter (other power quality disturbances are discussed in the next chapter).

"The lights are blinking" is the most common customer complaint to utilities. Other common complaints are "flickering," "clocks blinking," or "power out." The first step to improving power quality is identifying the actual problem. Sustained interruptions are the easiest to classify since the power is usually out when the customer calls. The "blinking" is harder to classify:

- Is it momentary interruptions caused by faults on the feeder serving the customer?
- Is it voltage sags caused by faults on lateral taps or adjacent feeders?
- Is it periodic voltage flicker caused by an arc welder or some other fluctuating load on the same circuit?

Some strategies for identifying the problem are:

- For commercial or industrial customers, does the customer lose all computers or just some of them? Losing all indicates the problem is momentaries; losing some indicates the problem is sags.
- Is it just the lights flickering? Do any computers or other electronic equipment reboot or reset? If it is just the lights, the problem is likely to be voltage flicker caused by some fluctuating load, which could be in the facility that is having problems.
- If approximate times of events are available from the customer, we can compare these times against the times of utility protective device operations. Of course to do this, the utility times must be recorded by a SCADA system or a digital relay or recloser controller. If these are available, it is often possible to correlate a customer outage to a utility protective device. If the protective device is a circuit breaker or recloser upstream of the customer, the cause was probably a momentary interruption. If the protective device is on an adjacent circuit or the subtransmission system, the likely cause was a voltage sag.
- A review of the number of operations of the protective devices on the circuit, if these records are kept, can reveal whether the customer is seeing an abnormal number of momentary interruptions or possibly sags from faults on adjacent feeders.
- Does the flickering occur because of changes in the customer load? For example in a house, does sump-pump starting cause the lights to dim in another room? If so, look for a local problem. A likely candidate — a loose neutral connection — causes a reference shift when load is turned on or off.
- Are other customers on the circuit having problems? If so, then the problem is probably due to momentary interruptions and not just a customer that is very sensitive to sags. Momentary interruptions affect most end users; voltage sags only impact the more sensitive end users.

2.1 Location

Fault location is the primary factor that determines the disturbance severity to customers. Figure 2.1 shows several fault locations and how they impact a specific customer differently. A fault on the mains causes an interruption for the customer. If the fault is permanent, the customer has a long-duration interruption, but if the fault is temporary, the interruption is short as the protective device recloses successfully. A fault on a lateral tap causes a voltage sag unless fuse saving is used. With fuse saving, the fault on the tap

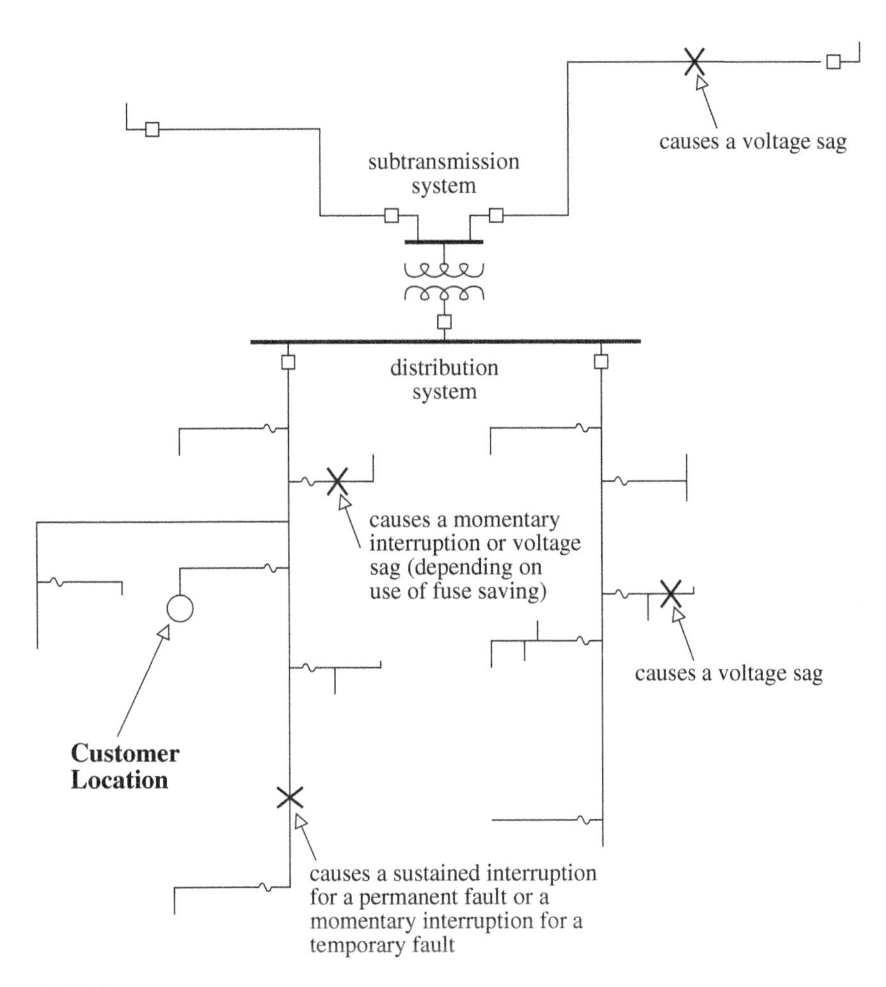

FIGURE 2.1
Example distribution system showing fault locations and their impact on one customer.

causes a momentary interruption as the substation breaker or recloser tries to prevent the fuse from blowing.

Faults on adjacent feeders cause voltage sags, the duration of which depends on the clearing time of the protective device. The depth of the sag depends on how close the customer is to the fault and the available fault current. Faults on the transmission system cause sags to all customers off of nearby distribution substations. We can depict all of the possible fault locations by areas of exposure or areas of vulnerability as shown in Figure 2.2. Each exposure area defines the vulnerability for the specific customer. For sags, we have different areas of vulnerability based on the severity of the sag. An outline of the area that causes sags to below 50% is tighter than the area of vulnerability for sags to below 70%. We can use the area of vulner-

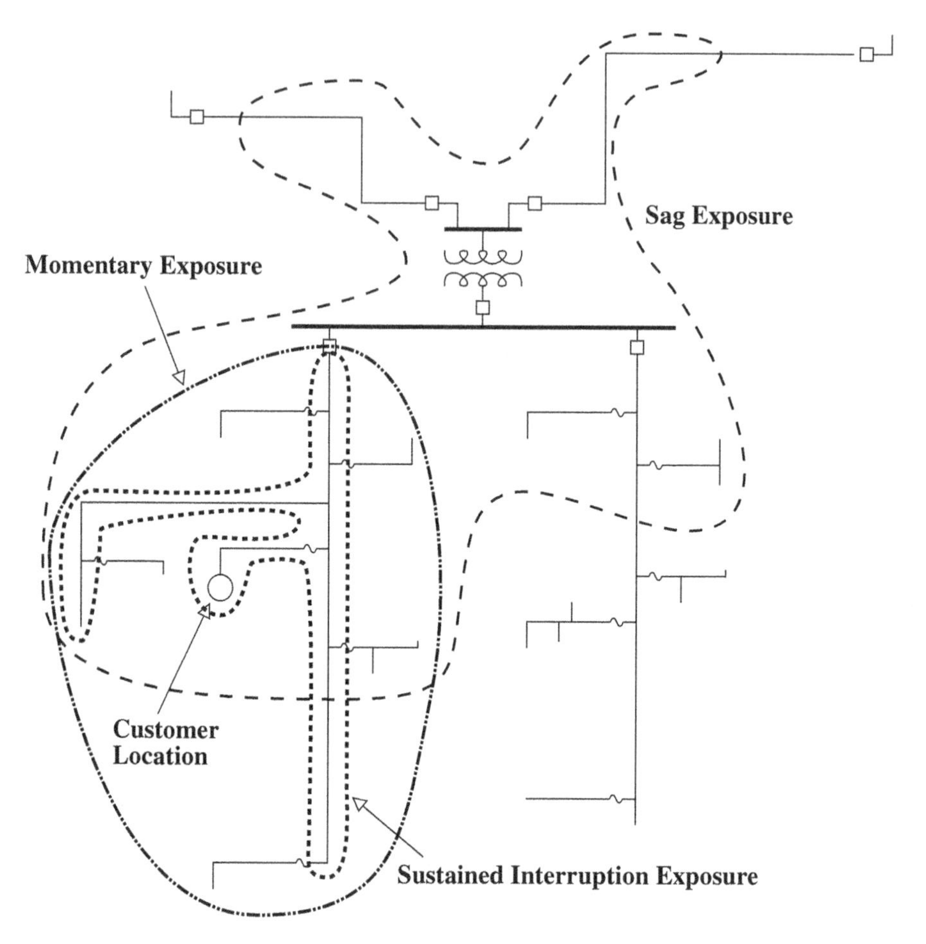

FIGURE 2.2
Example distribution system showing outlines of circuit exposure that cause a voltage sag, a momentary interruption, and a sustained interruption for one customer location.

ability curves to help target maintenance and improvements for important sensitive customers.

2.2 Momentary Interruptions

Momentary interruptions primarily result from reclosers or reclosing circuit breakers attempting to clear temporary faults, first opening and then reclosing after a short delay. The devices are usually on the distribution system; but at some locations, momentary interruptions also occur for faults on the subtransmission system. Terms for short-duration interruptions include *short interruptions, momentary interruptions, instantaneous interruptions,* and

TABLE 2.1

Surveys of MAIFI

Survey	Median
1995 IEEE (IEEE Std. 1366-2000)	5.42
1998 EEI (EEI, 1999)	5.36
2000 CEA (CEA, 2001)	4.0

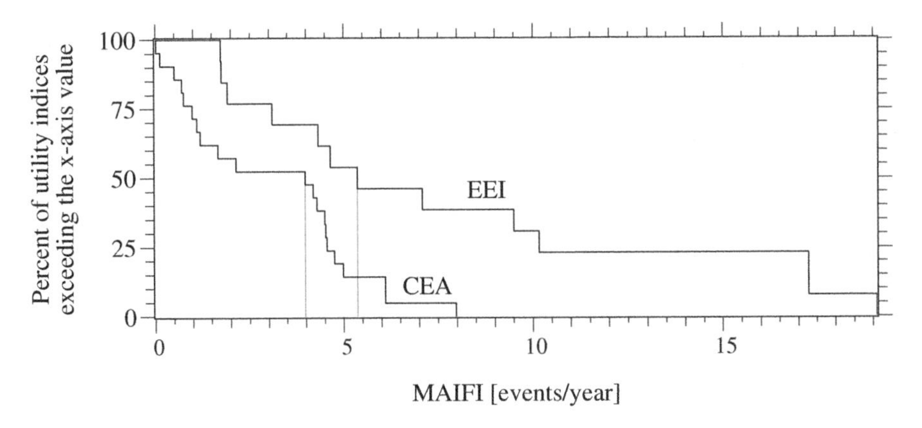

FIGURE 2.3
Distribution of utility MAIFI indices based on industry surveys by EEI and CEA. (Data from [CEA, 2001; EEI, 1999].)

transient interruptions, all of which are used with more or less the same meaning. The dividing line for duration between sustained and momentary interruptions is most commonly thought of as 5 min (1 min is also a common definition).

Table 2.1 shows the number of momentary interruptions based on surveys of the reliability index MAIFI. MAIFI is the same as SAIFI, but it is for short-duration rather than long-duration interruptions.

The number of momentary interruptions varies considerably from circuit to circuit and utility to utility. For example, in the EEI survey, the median of the utility averages is 5.4, but MAIFI ranged from 1.4 at the "best" utility to 19.1 at the "worst." Weather is obviously an important factor but so are exposure and utility practices. See Figure 2.3 for distributions of utility survey results.

There is a difference between the reliability definition and the power quality definition of a momentary interruption. The reliability definition (IEEE Std. 1366-2000) is

> A single operation of an interrupting device that results in zero voltage.

In addition, there is a distinction (IEEE Std. 1366-2000) between momentary interruptions and momentary interruption *events*:

> An interruption of duration limited to the period required to restore service by an interrupting device. Note: Such switching operations must be completed in a specified time not to exceed 5 min. This definition includes all reclosing operations which occur within 5 min of the first interruption. For example, if a recloser or breaker operates two, three, or four times and then holds, the event shall be considered one momentary interruption event.

Momentary interruption events and the associated index $MAIFI_E$ (E for event) better represent the impact on customers. Since we expect the first momentary disrupts the device or process, subsequent interruptions are unimportant. Momentary interruptions are most commonly tracked by using breaker and recloser counts, which implies that most counts of momentaries are based on MAIFI and not $MAIFI_E$. To accurately count $MAIFI_E$, a utility must have a SCADA system or other time-tagging recording equipment.

The power quality definition of a momentary interruption (IEEE Std. 1159-1995) is based on the voltage characteristics rather than the cause:

> A type of *short duration variation*. The complete loss of voltage (<0.1 pu) on one or more phases for a time period between 0.5 cycles and 3 sec.

Several extra events fall under the power quality definition of a momentary interruption. The power quality definition includes both operations of interrupting devices as well as very deep voltage sags. For this book, the reliability definition of a momentary interruption is used. The difference is worth remembering. Momentary interruptions that are tracked by using breaker and recloser counts are different from momentary interruptions recorded by power quality recorders. Table 2.2 shows momentary interruptions as recorded by several power quality studies using the power quality definition and an estimate of the reliability definition where very short events are excluded.

TABLE 2.2

Average Annual Number of Momentary Interruptions from Monitoring Studies

Study	Power Quality Definition[a] 1 cycle–10 sec	Reliability Definition[a] 20 cycles–10 sec
EPRI feeder sites (5-min filter)	6.4	4.5
NPL (5-min filter)	7.9	6.8
CEA primary (no filter)	3.2	1.3
CEA secondary (no filter)	6.5	2.8

[a] These are not industry standard definitions, just arbitrary time windows chosen to illustrate that the power quality definition of momentary interruptions has more events than a reliability definition.

Source: Dorr, D. S., Hughes, M. B., Gruzs, T. M., Juewicz, R. E., and McClaine, J. L., "Interpreting Recent Power Quality Surveys to Define the Electrical Environment," *IEEE Transactions on Industry Applications*, vol. 33, no. 6, pp. 1480–7, November 1997.

Momentaries can be improved in several ways including the following:

- *Reduce faults* — Tree trimming, tree wire, animal guards, arresters, circuit patrols, etc.
- *Reclose faster*
- *Limit the number of customers interrupted* — Single-phase reclosers, extra downstream reclosers, not using fuse saving, etc.

2.3 Voltage Sags

Voltage sags cause some of the most common and hard-to-solve power quality problems. Sags can be caused by faults some distance from a customer's location. The same voltage sag affects different customers and different equipment differently. Solutions include improving the ride-through capability of equipment, adding additional protective equipment (such as an uninterruptible power supply), or making improvements or changes in the power system.

A voltage sag is defined as an rms reduction in the ac voltage, at the power frequency, for durations from a half cycle to a few seconds (IEEE Std. 1159-1995). Sags are also called *dips* (the preferred European term). Faults in the utility transmission or distribution system cause most sags. Utility system protective devices clear most faults, so the duration of the voltage sag is the clearing time of the protective device.

Voltage-sag problems are a contentious issue between customers and utilities. Customers report that the problems are due to events on the power system (true), and that they are the utility's responsibility. The utility responds that the customer has overly-sensitive equipment, and the power system can never be designed to be disturbance free. Utilities, customers, and the manufacturers of equipment all share some of the responsibility for voltage sag problems. There are almost no industry standards or regulations to govern these disputes, and most are worked out in negotiations between a customer and the utility.

Terminology is a source of confusion. A 30% voltage sag can be interpreted as the voltage dropping to 70% of nominal or to 30% of nominal. Be more precise and say a "sag *to* X (volts or percent)." There is also some difference between a sag to 60% of nominal and a sag to 60% of the prefault voltage. Since most (but not all) equipment is sensitive to the actual voltage, generally refer to sags based on the percentage of nominal voltage.

Figure 2.4 shows a voltage sag that caused the system voltage to fall to approximately 45% of nominal voltage for 4.5 cycles.

Voltage sags can be improved with several methods on the utility system:

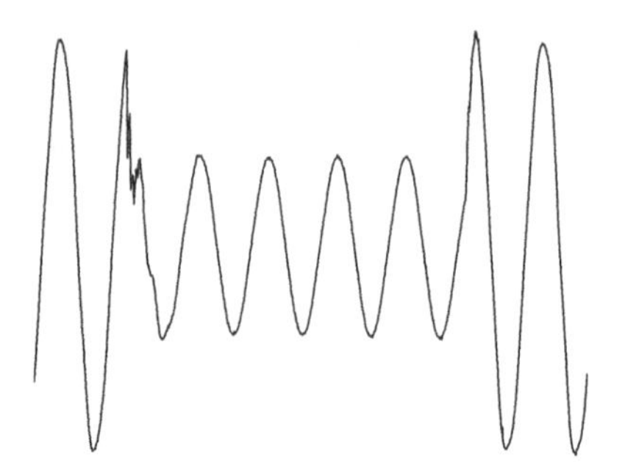

FIGURE 2.4
Example voltage sag caused by a fault.

- *Reduce faults* — Tree trimming, tree wire, animal guards, arresters, circuit patrols
- *Trip faster* — Smaller fuses, instantaneous trip, faster transmission relays
- *Support voltage during faults* — Raising the nominal voltage, current-limiting fuses, larger station transformers, line reactors

The voltage during the fault at the substation bus is given by the voltage-divider expression in Figure 2.5 based on the source impedance (Z_s), the feeder line impedance (Z_f), and the prefault voltage (V).

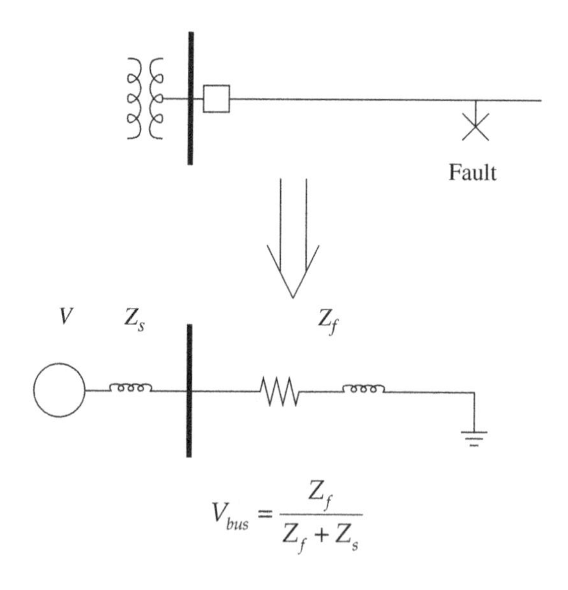

FIGURE 2.5
Voltage divider equation giving the voltage at the bus for a fault downstream. (This can be the substation bus or another location on the power system.)

The voltage sags deeper for faults electrically closer to the bus (smaller Z_f). Also, as the available fault current decreases (larger Z_s), the sag becomes deeper. The source impedance includes the transformer impedance plus the subtransmission source impedance (often, subtransmission impedance is small enough to be ignored). The impedances used in the equation depend on the type of fault it is. For a three-phase fault (giving the most severe voltage sag), use the positive-sequence impedance ($Z_f = Z_{f1}$). For a line-to-ground fault (the least severe voltage sag), use the loop impedance, which is $Z_f = (2Z_{f1} + Z_{f0})/3$. A good approximation is one ohm for the substation transformer (which represents a 7 to 8-kA bus fault current) and 1 Ω/mi (0.6 Ω/km) of overhead line for ground faults. For accuracy, use complex division since the impedances are complex, but for back-of-the-envelope, first-approximation calculations, use the impedance magnitude.

Another way to approximate the voltage divider equation is to use the available short-circuit current at the substation bus and the available short-circuit current at the fault location:

$$V_{bus} = 1 - \frac{I_f}{I_s}$$

where

V_{bus} = per unit voltage at the substation
I_f = the available fault current on the feeder at the fault location
I_s = the available fault current at the substation bus

Note that this can be used for any type of fault as long as the appropriate fault values are used in the equation. If the angles are ignored, the equation is an approximation (which is usually acceptable). Figure 2.6 shows a profile of the substation bus voltage for faults at the given distance along the line for 12.47, 24.94, and 34.5 kV. The higher-voltage systems have more severe voltage sags for faults at a given distance. The graph also shows that three-phase faults cause more severe sags. Figure 2.7 compares sags on underground and overhead systems.

The effect of feeder faults on voltage sags at the substation bus can be estimated with the following equation:

$$S(V_{sag}) = n_f \lambda \frac{V_{sag}}{1 - V_{sag}} \left(\frac{Z_s}{Z_f} \right)$$

where

S = annual number of sags per year where the voltage sags below V_{sag}
V_{sag} = per unit voltage sag level of interest (in the range of 0 to 1, e.g., 0.7)
n_f = number of feeders off of the bus

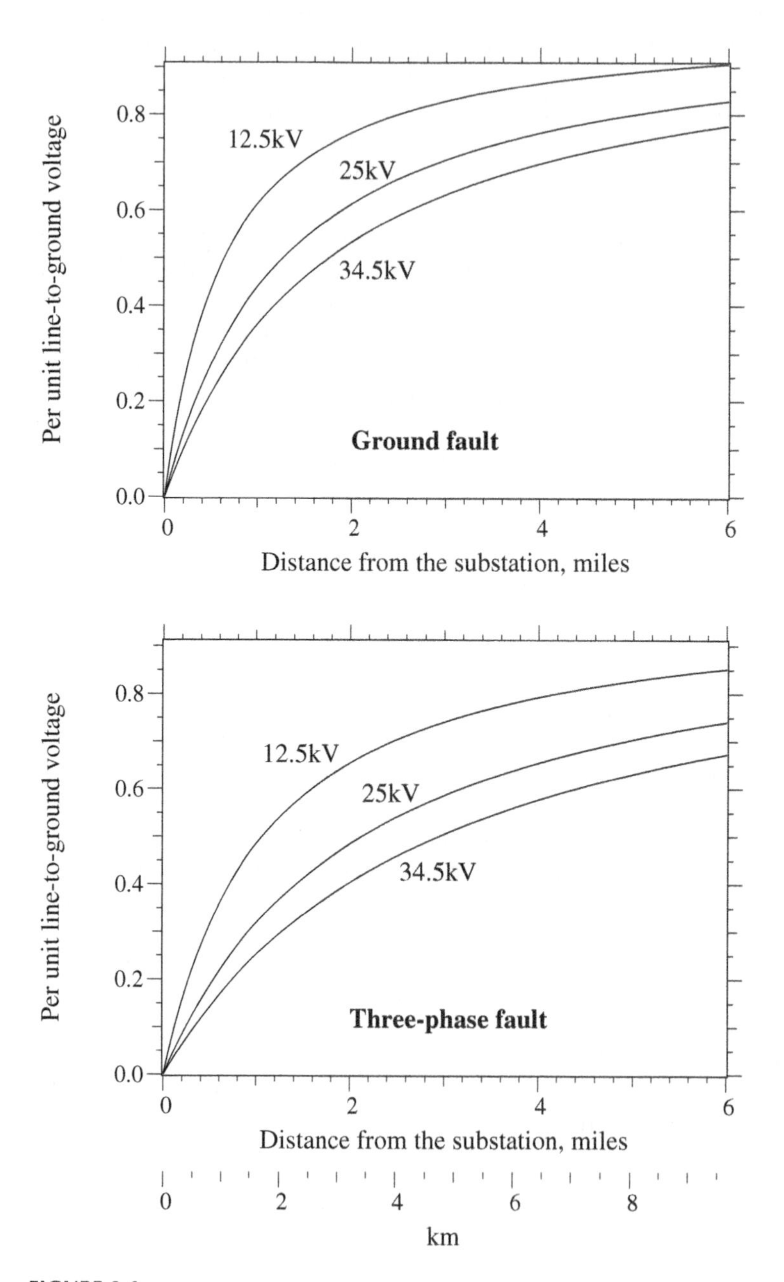

FIGURE 2.6
Substation voltage profile for faults at the given distance (single-phase and three-phase faults are shown for each voltage — the circuit parameters for the 500-kcmil circuit are the same as those in Figure 4.11).

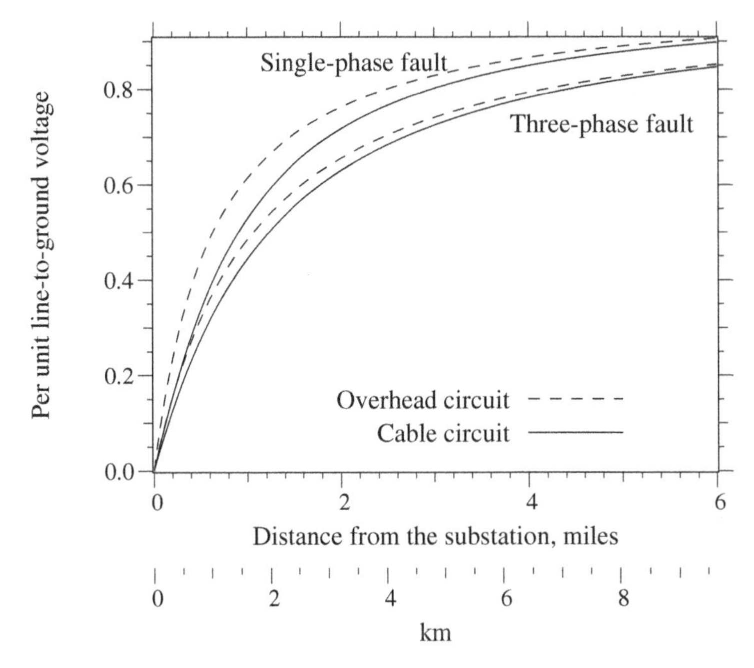

FIGURE 2.7
Comparison of substation voltage for faults on overhead circuits and cable circuits at the given distance (single-phase and three-phase faults are shown; the circuit parameters are the same as those in Figure 4.11 and Figure 4.12).

λ = feeder mains fault rate per mile (or other unit of distance) per phase including faults on laterals and including both temporary and permanent faults

Z_f = feeder impedance, Ω/mi (or other unit of distance); usually use $Z_f = (2Z_1+Z_0)/3$ for ground faults

Z_s = source impedance, Ω

The distribution of voltage sags based on this equation is shown in Figure 2.8 for some common parameters. Several points are noted from this analysis on voltage sags:

- *Exposure* — For 15-kV circuits, we can ignore exposure beyond the first 2 or 3 mi (4 or 5 km) for sags to the bus voltage. The first mile or two is most important as far as circuit improvement, maintenance, or application of current-limiting fuses.
- *System voltage* — Sags are more severe on higher voltage distribution systems (especially at 34.5 kV). A fault 4 mi from the substation sags the voltage much more on a 25-kV system than on a 12-kV system because the substation transformer is a higher impedance relative to the line impedance at higher system voltages. For 24.94 kV, exposure as far as 5 mi from the station is significant.

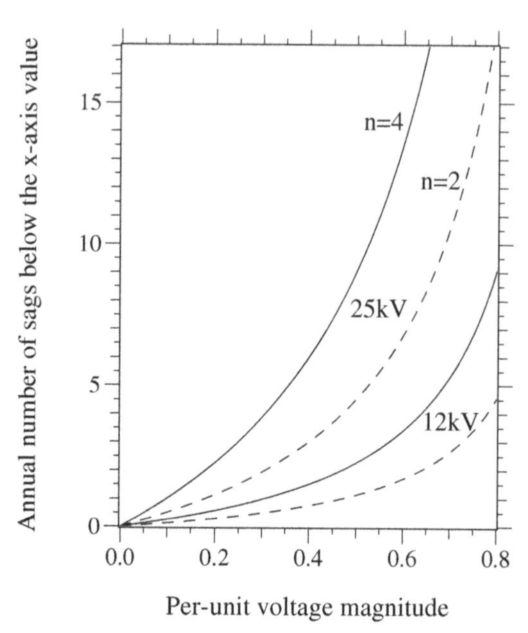

FIGURE 2.8
Cumulative distribution of substation bus voltage sags per year for the given (25-MVA, 10% transformer, 500-kcmil feeder, n = 2 or 4 feeders off of the bus, λ = 1 faults/phase/mile of mains/year, assumes line-to-ground faults only). Copyright © 2003. Electric Power Research Institute. 1001665. *Power Quality Improvement Methodology for Wires Companies.* Reprinted with permission.

- *Single vs. three-phase faults* — Three-phase faults cause more severe sags than single-line-to-ground faults. Three-phase faults farther away can pull the voltage down.
- *Underground vs. overhead* — All-underground circuits have more exposure to sags because cables have lower impedance than overhead lines.
- *Number of feeders* — The number of sags on the station bus is directly proportional to the number of feeders off the bus.
- *Transformer impedance* — A lower station transformer impedance (a bigger transformer or lower percent impedance) improves voltage sags.
- *Bus tie* — It does not matter whether a substation bus tie is open or closed. If it is open, a fault only affects half of the feeders. A fault that does occur forces a deeper sag because of a higher effective source impedance. These two effects tend to cancel each other.
- *Voltage regulation* — Raising the nominal voltage improves the voltage seen by customers during a fault. Say that a fault drops the voltage to 0.8 per unit, and the prefault voltage was 1.0 per unit. If the prefault voltage were 1.1 per unit, the voltage during the sag is 0.88 per unit. This is not a big difference, but for equipment sensitive

TABLE 2.3

Line-to-Ground and Line-to-Line Voltages on the Low-Voltage Side of a Transformer with One Phase on the High-Voltage Side Sagged to Zero

Voltages	Primary Voltages			Voltages Downstream of a Delta–Wye Transformer		
Line-ground	0.00	1.00	1.00	0.58	0.58	1.00
Line-line	0.58	0.58	1.00	0.33	0.88	0.88

to sags to 0.7 to 0.85 per unit, higher voltages appreciably reduce the number of tripouts.

Customers at the end of a circuit have more severe voltage sags because almost all faults upstream appear as little or no voltage (most actually fit the power quality definition of an interruption, a voltage to below 10%).

2.3.1 Effect of Phases

Three-phase loads are often controlled by single-phase devices (the controls are often the most sensitive element). The effect on three-phase customers depends on how loads are connected and depends on the transformer connection as shown in Table 2.3. In general, if the transformer causes more phases to be affected, the voltage drop is less severe. One situation is not always better than the other. Severity depends on which phases the sensitive devices are located. The type and design of the device and its controls are also factors.

For facility equipment connected *line-to-line*, the wye – wye transformer connection provides the best performance. For facility equipment connected *line-to-ground*, the delta – wye facility transformer is best.

Single-phase sags on distribution systems are more common than two- or three-phase sags. This is expected since most faults on distribution systems are single phase. For example, in EPRI's Distribution Power Quality Study (EPRI TR-106294-V2, 1996), about 64% of voltage sags to below 70% were single phase, while three-phase sags made up 25%, and two-phase sags, 10%. For severe sags below 30% voltage, three-phase events are more common; more than half are three-phase events (see Figure 2.9). This includes momentary interruptions, most of which are three-phase.

2.3.2 Load Response

During a voltage sag, rotating machinery supports the voltage by feeding current back into the system. Synchronous motors and generators provide the largest boost. Induction motors also provide benefit, but the support decays quickly.

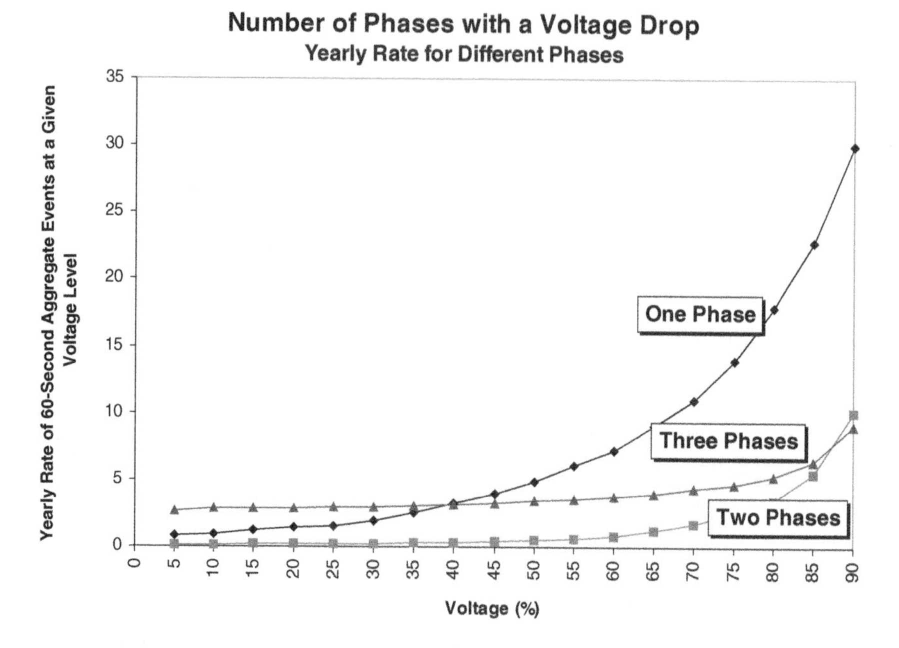

FIGURE 2.9
Rate of number of phases with a voltage drop in the EPRI DPQ study. (Copyright © 1996. Electric Power Research Institute. TR-106294-V2. *An Assessment of Distribution System Power Quality: Volume 2: Statistical Summary Report.* Reprinted with permission.

Increasing a motor's inertia is one way to increase the ride through of the motor, which also increases the support to other loads in the facility.

Following a sag, however, the response of loads — particularly motors — may further disturb the voltage. During a sag, motors slow down. After the sag, the motors draw inrush current to speed up. If motors are a large enough portion of the load, this inrush pulls the voltage down, delaying the recovery of voltage. Motors with small slip and those with large inertia draw the most inrush following a sag. These effects are more severe for customers or areas with a large percentage of motor loads and for longer fault clearing times (Bollen, 2000; IEEE Std. 493-1997).

An extreme case of motor inrush sometimes happens with air conditioners. Single-phase air conditioner compressors are prone to stall during voltage sags; during which, the compressor draws locked rotor current, about five or six times normal. Tests by Williams et al. (1992) found that voltages below 60% of nominal for five cycles stalled single-phase air conditioners. Longer-duration sags also stall compressors for less severe sags (in the range of 60 to 70% of nominal). The compressor stays stalled long after the system voltage has returned to normal. It keeps drawing current until thermal overload devices trip the unit, which can take one half of a second. On the distribution system, this extra current may trip breakers or blow fuses in addition to aggravating the voltage sag.

Adjustable-speed drives and other loads with capacitors (mainly rectifiers) also draw inrush following a voltage sag. During the sag, rectifiers stop drawing current until the dc voltage on the rectifier drops to the sagged voltage. After the sag, the rectifier draws inrush to charge the capacitor. This spikes to several times normal, but the duration is short relative to motor inrush. The inrush may blow fuses or damage sensitive electronics in the rectifier. For severe sags, much of the rectifier-based load trips off, which reduces the inrush.

Normally, we neglect the load response for voltage sag evaluations, but occasionally, we must consider the response of the load, either for its direct impact on voltage sags, or for the impact of the inrush.

2.3.3 Analysis of Voltage Sags

The calculation of the voltage magnitude at various points on a system during a fault at a given location is easily done with any short-circuit program. We make the fairly accurate assumption that the fault impedance is zero. The engineer or computer program finds the duration of the sag using the time-current characteristics of the protective device that should operate along with the fault current through it.

Based on a short-circuit program, the *fault positions method* repeatedly applies faults at various locations and tallies the voltages at specified locations during the faults. The procedures, which may apply thousands of fault locations, result in predictions of the number of voltage sags below a given magnitude at the specified locations. This procedure is well documented in the *Gold Book* (IEEE Std. 493-1997) [see also (Conrad et al., 1991)].

The faults are applied along each line in a system. The end results are scaled by the fault rate on the line, which can be based on historical results or typical values for the voltage and construction.

We need considerable detail for the fault-positions analysis, especially a complete system model including proper zero-sequence impedances and transformer connections (these are left out of many transmission system load-flow models).

Another simpler method for voltage sags is the *method of critical distances* (Bollen, 2000). The approach is to find the farthest distance, the *critical* distance, to a fault that causes a sag of a given magnitude. Pick a sag voltage of interest, 0.7 per unit for example. Find the critical distance for the chosen voltage. Using a feeder map, add up the circuit lengths within the critical distance. Multiply the total exposed length by the fault rate — this is the number of events expected. This method is not as accurate as the fault positions method, but is much simpler: we can calculate the results by hand, and the process of doing the calculations provides insight on the portions of distribution and transmission system that can cause sags to the given customer. We can also target this *area of vulnerability* for inspection or additional maintenance or apply faster protection schemes covering those circuits (to clear faults and sags more quickly).

2.4 Characterizing Sags and Momentaries

2.4.1 Industry Standards

The most commonly cited industry standard for ride through was developed by the Information Technology Industry Council (ITI) (Figure 2.10). The ITI curve updates the CBEMA curve (Computer Business Equipment Manufacturers' Association, which became ITI) and is often referred to as the new CBEMA curve. The ITI curve is not an actual tested standard — computers do not have to be certified to pass some test. The ITI curve is used as a

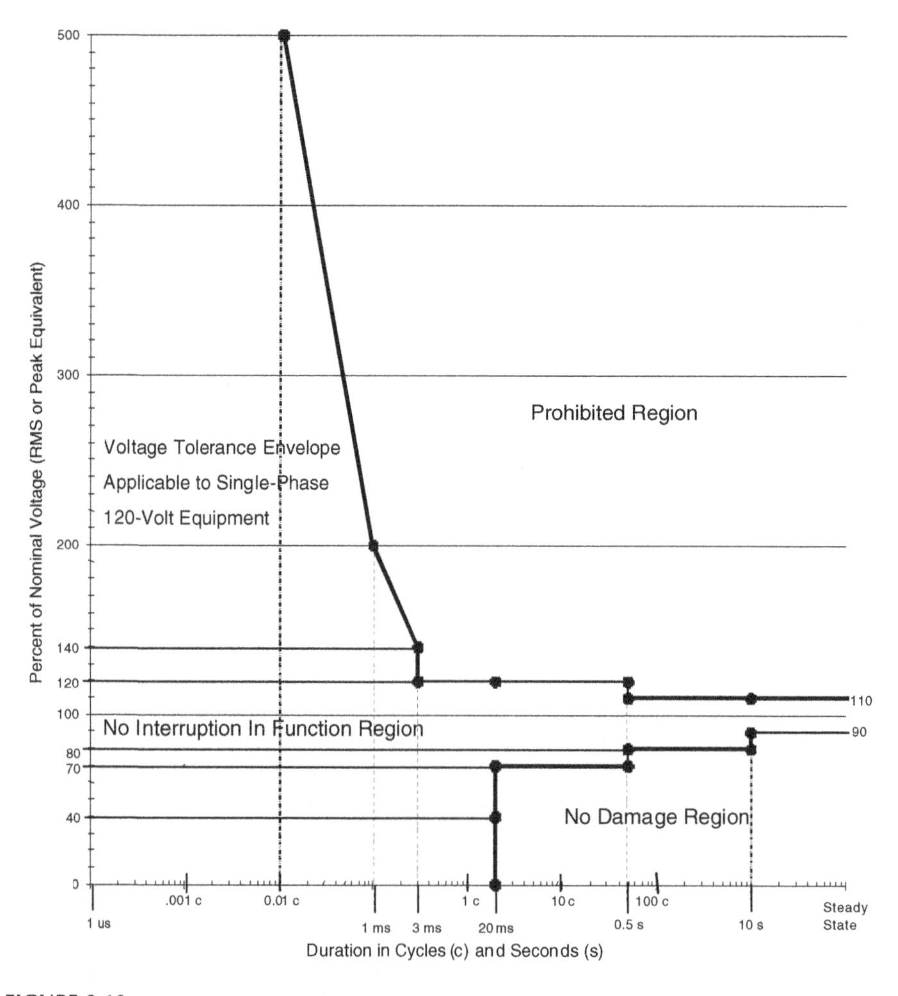

FIGURE 2.10
ITI curve that shows the *typical* voltage sensitivity of information technology equipment. (From [Information Technology Industry Council (ITI), 2000]. With permission.)

benchmark indicator for comparison of power quality between sites and to track performance over time. Because the ITI curve somewhat represents the ride through of computers, we can single out events below the ITI curve as "suspects," which may trip sensitive equipment.

Another major equipment standard has been produced by the semiconductor industry (SEMI F47-0200, 1999). The major advance of the SEMI set of standards is that there is an actual test standard for the equipment. To meet the SEMI standard, equipment must pass a series of voltage sag tests (SEMI F42-0600, 1999). The standard defines many factors, including sag generator and other test apparatus requirements, sampling specimens, test procedure, and reporting of test results. The SEMI standard is only for single-phase sags; for three-phase equipment with a neutral, six tests are done: each phase-to-neutral voltage is "sagged," and each phase-to-phase voltage is sagged in turn. For three-phase equipment without a neutral, each phase-to-phase voltage is tested with a sag generator.

The SEMI curve focuses exclusively on voltage sags. In some cases, the SEMI curve is more strict than the ITI curve, and it appears that way when the two curves are graphed together as in Figure 2.11. The SEMI curve has a deeper voltage sag characteristic. The most severe point on the SEMI curve is the 0.2-sec sag for a voltage to 50% of nominal. However, some equipment could meet the SEMI requirement but not pass the ITI curve points. The main types of equipment that fall into this category are relays and contactors. The ITI curve has a 0.02-sec interruption that is enough to disengage many relays and contactors that may survive a 0.2-sec sag to 50% of nominal voltage.

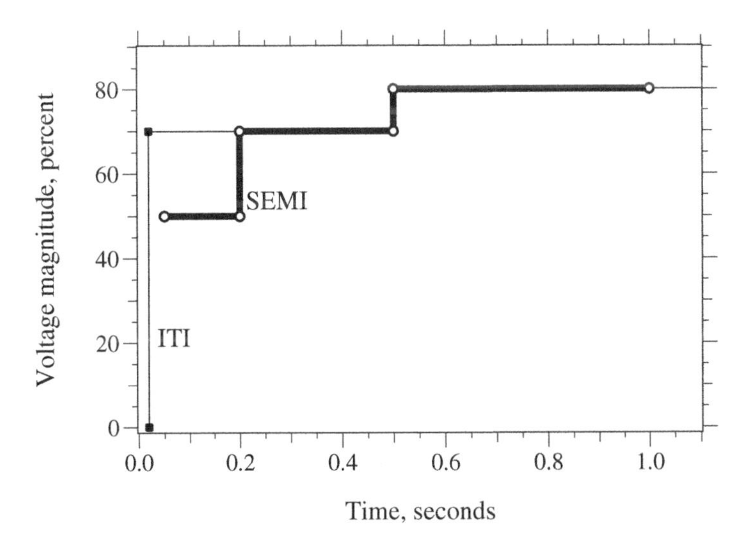

FIGURE 2.11
SEMI voltage sag ride-through requirement compared against the ITI curve. (SEMI curve from [SEMI F47-0200, 1999].)

Several power quality indices have been introduced that are similar to the reliability indices (EPRI TP-113781, 1999). Utilities can use these for some of the same purposes as reliability indices: targeting areas for maintenance and circuit upgrades, tracking the performance of regions, and documenting performance to regulators. The most widely used index is SARFI (EPRI TP-113781, 1999; Sabin et al., 1999) defined as

> $SARFI_X$, System Average RMS (Variation) Frequency Index: $SARFI_X$ represents the average number of specified rms variation measurement events that occurred over the assessment period per customer served, where the specified disturbances are those with a magnitude less than X for sags or a magnitude greater than X for swells.

$$SARFI_X = \frac{\sum N_i}{N_T}$$

where

X = rms voltage threshold; possible values — 140, 120, 110, 90, 80, 70, 50, and 10

N_i = number of customers experiencing short-duration voltage deviations with magnitudes above X% for X > 100 or below X% for X < 100 due to measurement event *i*

N_T = number of customers served from the section of the system to be assessed

The breakpoints were not chosen arbitrarily. The 90, 80, and 70% thresholds are boundaries of the ITI curve, the 50% threshold is a typical breakpoint for motor contactors, and 10% is the dividing line between a sag and an interruption. Two special variations of SARFI have also been defined. $SARFI_{ITIC}$ is the number of events below the lower ITI curve. In similar fashion, $SARFI_{SEMI}$ is the number of events below the SEMI curve. SARFI can be applied for one monitor (and one customer) or for several monitored locations. It is difficult to extend this concept to make SARFI a system-wide performance indicator like SAIFI — what is straightforward for reliability indices becomes much more complicated for sags because a fault causes different voltages at different locations on the distribution system. It is difficult to find a system-wide average without a vast number of monitors. Approximations must be used to estimate the effects at different customers based on a small number of monitored points.

2.4.2 Characterization Details

Several disturbances often occur within a short time of each other. Commonly, a breaker or recloser goes through several reclosing attempts. The customer sees a sequence of voltage sags. If one of these events causes an end-use disruption, from their point of view, it does not matter if additional events follow within the next few minutes, as the customer is already disturbed. To account for this, we can aggregate events within a rolling time window. Commonly, time windows are 1 and 5 min for calculating $SARFI_X$ or other power quality benchmarks.

Since voltage sags can have different impacts on each phase, how do we account for the differences between a three-phase sag and a single-phase sag? We can tabulate sags in two different ways:

- *Per phase* — Each phase is tracked independently. A three-phase sag counts three times that of a single-phase sag. Single-phase recorders automatically calculate the number of sags per phase.
- *Minimum phase* — A sag event is recorded as the lowest of the three phase voltages. A three-phase sag counts the same as a single-phase sag. $SARFI_X$ uses this approach.

Both approaches are useful depending on the customer and load characteristics. The per-phase method is better for single-phase customers and for customers with three-phase load that is more sensitive to multiple-phase sags. The minimum-phase method is better for facilities where sags on any of the three phases could trip a process. At a three-phase location, the minimum-phase method gives higher numbers of voltage sags.

The line-to-ground and line-to-line voltages may be significantly different during a voltage sag. Ideally, we want to record and benchmark what the critical load sees, but sometimes that is unknown (and, some facilities may have critical loads connected line to line and line to ground). Normally, SARFI is tracked based on however the recorders are connected.

Most voltage sags have a simple shape — the voltage drops in magnitude and stays at a constant value until the fault clears. After that, the voltage returns to its pre-sag value. The rms change is approximately a rectangular wave. The rectangular shape makes classification easy — only a magnitude and a duration are needed. Sometimes, sags do not follow the rectangular shape. If the fault current is not constant, the voltage will not be constant. If the fault evolves from a single line-to-ground fault into a multiple-phase fault, the voltage will change. These types of events are hard to classify, but most of the time, we can ignore them for the purposes of monitoring and collecting statistics at a site. For analysis of specific events that disrupted equipment, review of the rms shape may provide additional meaning beyond just having a magnitude and a duration.

2.5 Occurrences of Voltage Sags

Several power quality monitoring studies have characterized the frequency of voltage sags. The two most widely quoted studies are EPRI's Distribu-

TABLE 2.4

Average Annual Number of Voltage Sags below the Given Magnitude for Longer than the Given Duration from the NPL Data with a Five-Minute Filter

	Duration							
Magnitude	1 Cycle	6 Cycles	10 Cycles	20 Cycles	0.5 sec	1 sec	2 sec	10 sec
87%	126.4	56.8	36.4	27.0	23.0	18.1	14.5	5.2
80%	44.8	23.7	17.0	13.9	12.2	10.0	8.0	4.3
70%	23.1	17.3	14.5	12.8	11.5	9.7	7.9	4.3
50%	15.9	14.1	12.9	11.8	10.6	9.4	7.8	4.3
10%	12.2	12.0	11.7	11.0	10.2	9.0	7.5	4.2

Source: Dorr, D. S., Hughes, M. B., Gruzs, T. M., Juewicz, R. E., and McClaine, J. L., "Interpreting Recent Power Quality Surveys to Define the Electrical Environment," *IEEE Transactions on Industry Applications*, vol. 33, no. 6, pp. 1480–7, November 1997.

tion Power Quality (DPQ) study and the National Power Laboratory's end-use study.

NPL's end-use study recorded power quality at the point of use at residential, commercial, and industrial customers. At 130 sites within the continental U.S. and Canada, single-phase line-to-neutral monitors were connected at standard wall receptacles (Dorr, 1995). The survey resulted in a total of 1200-monitor months of data. Table 2.4 shows the average number of voltage sags that dropped below the given magnitude for longer than the given duration.

EPRI's Distribution Power Quality (DPQ) project recorded power quality in distribution substations and on distribution feeders, measured on the primary at voltages from 4.16 to 34.5 kV (EPRI TR-106294-V2, 1996; EPRI TR-106294-V3, 1996). It was seen that 277 sites resulted in 5691 monitor-months of data. In most cases three monitors were installed for each randomly selected feeder, one at the substation and two at randomly selected places along the feeder. Table 2.5 shows average numbers of voltage sags for a given magnitude and duration for the DPQ data.

TABLE 2.5

Average Annual Number of Voltage Sags Below the Given Magnitude for Longer than the Given Duration from the EPRI Feeder Data with a Five-Minute Filter

	Duration							
Magnitude	1 Cycle	6 Cycles	10 Cycles	20 Cycles	0.5 sec	1 sec	2 sec	10 sec
90%	77.7	31.2	19.7	13.5	10.7	7.4	5.4	1.8
80%	36.3	17.4	12.4	9.3	7.9	6.4	4.9	1.7
70%	23.9	13.1	10.3	8.3	7.2	6.2	4.8	1.7
50%	14.6	9.5	8.4	7.5	6.6	5.9	4.6	1.7
10%	8.1	6.5	6.4	6.2	5.6	5.1	4.0	1.7

Source: Dorr, D. S., Hughes, M. B., Gruzs, T. M., Juewicz, R. E., and McClaine, J. L., "Interpreting Recent Power Quality Surveys to Define the Electrical Environment," *IEEE Transactions on Industry Applications*, vol. 33, no. 6, pp. 1480–7, November 1997.

TABLE 2.6

Annual Number of Power Quality Events (Upper Quartile, Median, and Lower Quartile) for the EPRI DPQ Feeder Sites with a One-Minute Filter

	Duration, seconds						
Voltage	0	0.02	0.05	0.1	0.2	0.5	1
0.9	32.8 57.5 104.8	30.8 49.0 95.1	24.4 35.3 65.6	13.6 22.7 38.7	7.6 13.2 24.0	3.3 7.3 14.2	1.4 3.2 8.9
0.8	16.4 31.6 54.1	14.8 26.0 50.1	12.1 20.9 37.9	8.1 15.0 25.1	4.9 9.6 16.9	2.4 5.3 11.0	0.9 2.7 7.5
0.7	10.1 20.5 33.8	8.6 18.8 32.7	8.1 15.3 27.6	5.8 11.3 18.8	4.0 7.8 13.5	1.8 4.5 9.3	0.9 2.5 7.0
0.5	4.7 9.7 19.2	4.5 9.0 17.4	4.2 7.7 14.3	3.5 5.9 11.2	2.3 5.0 9.6	1.4 3.3 7.7	0.8 2.2 5.7
0.3	2.1 4.8 12.8	1.8 4.5 11.0	1.6 4.2 9.5	1.4 3.6 8.6	1.1 3.5 8.3	0.8 2.8 6.6	0.5 1.6 5.1
0.1	0.9 3.2 8.3	0.9 2.9 7.8	0.8 2.8 7.8	0.8 2.7 7.8	0.7 2.7 7.8	0.5 2.2 6.1	0.3 1.6 4.9

Note: A B C represent the lower quartile *A*, the median *B*, and the upper quartile *C* of the total number of events below the given magnitude and longer than the given duration (up to 1 min).

As expected, the number of voltage sags is higher for the end-use NPL study than for the primary-level DPQ study. At the point of use, the nominal voltage is lower, which picks up more voltage sags, especially minor sags. End-use monitoring also picks up events caused internally, mainly voltage sags.

Table 2.6 shows cumulative numbers of voltage sags measured at sites during the DPQ study. Table 2.4 and Table 2.5 presented results based on averages — Table 2.6 shows the data based on the median, upper, and lower quartiles. One use of it is to estimate the number of times a year disturbances will affect a device — for example, if a device is sensitive to any event below a voltage of 50% of nominal for longer than 0.1 sec, then Table 2.6 predicts that at half of the sites in the U.S. distribution system, the device misoperates more than 5.9 times per year.

As an indicator, the average misrepresents the typical site power quality. The median represents site data better; here, by definition, 50% of sites have values higher than the median, and 50% have values lower. With balanced distributions such as the normal distribution, the average equals the median. In a skewed distribution, the average is higher than the median. Additionally, poor sites and anomalies such as a severe storm skew the average upward. In the DPQ data, the average is 31 to 115% higher than the median depending on the quality indicator as shown in Table 2.7.

2.5.1 Site Power Quality Variations

EPRI's Distribution Power Quality (DPQ) project allows us the opportunity to explore how power quality varies at different sites. Completed in 1995, the DPQ project collected data from 24 utility systems at a total of 277 locations on 100 distribution system feeders over a 27-month period. Site and circuit descriptors help us analyze the causes for site variations. Some

TABLE 2.7

Ratio of Median and Average for DPQ Site Statistics at Feeder Sites

	Median	Average	Ratio of Average to Median
$SARFI_{ITIC}$	21.27	27.86	131%
$SARFI_{SEMI}$	15.28	18.92	124%
$SARFI_{10}$	2.52	5.42	215%

notable details about the DPQ measurements and our analysis (Short et al., 2002):

- All measurements were on the distribution primary. Of course, most customers connect to the distribution secondary. Normally, this means that a customer's equipment sees more events below a given threshold. Also note that for three-phase customers, a delta – wye transformer distorts the secondary voltages relative to the primary voltages.
- All data was measured at three-phase points on the distribution circuit (single-phase locations were not monitored).
- We present all data based on the worst of the three phases, which is conservative because most faults are single phase. Single-phase customers see fewer sags. In addition, some three-phase equipment is less sensitive to single-phase sags than to three-phase sags.
- Most of the measurements are from phase to ground (the monitors on the ungrounded circuits show phase-to-phase measurements).
- We only used sites with at least 200 days of monitoring.

Power quality varies widely by site. Figure 2.12 shows cumulative distributions of different power quality indices along with statistics and a fit to a log-normal distribution. The left column (SARFI 70, 50, and 10) gives the average annual number of voltage sags below 70, 50, and 10%, which are most applicable for relays, contactors, and other devices that drop out quickly. $SARFI_X$ considers only short-duration rms events, defined as 1/2 cycle to one minute (IEEE Std. 1159-1995). The right column of Figure 2.12 shows data similar to the left column but for criteria that disregards very short events. The ITI curve (Information Technology Industry Council, 2000) disregards sags less than 0.02 sec, and the SEMI curve (SEMI F47-0200, 1999) disregards sags less than 0.05 sec. The indices that exclude short events are more appropriate for computer power supplies and other devices that ride through short-duration events. $SARFI_{10\ (>0.4sec)}$ is for momentary interruptions greater than 0.4 sec, which differentiates between deep sags and total loss of voltage due to operation of a breaker or recloser.

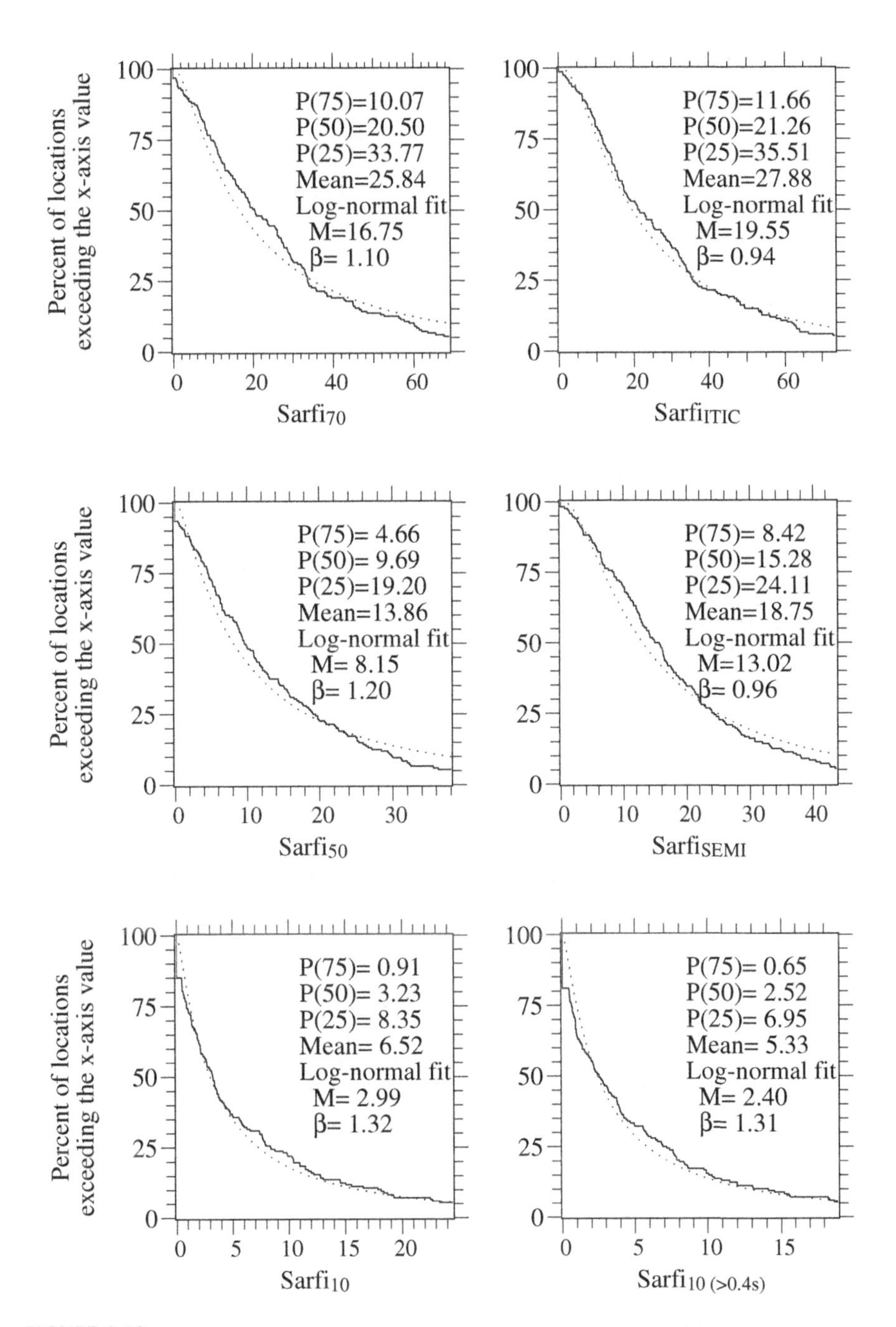

FIGURE 2.12
Cumulative distributions of DPQ feeder data along with statistics for various indices. SARFI 70, 50, and 10 gives the number of voltage sags below 70, 50, and 10%. $SARFI_{ITIC}$ and $SARFI_{SEMI}$ are events below the ITI curve and the SEMI curve, respectively. The dotted line fits a log-normal distribution.

TABLE 2.8

Statistics for Power Quality from the SEMI Monitoring Study, Which Are Primarily Transmission Service

		Median		
	Average	P(75%)	P(50%)	P(25%)
$SARFI_{ITIC}$	4.60	2.05	3.80	5.10
$SARFI_{SEMI}$	2.05	0.00	1.90	3.64
$SARFI_{70}$	4.40	2.05	3.50	5.10
$SARFI_{50}$	0.97	0.00	0.69	1.20
$SARFI_{10}$	0.24	0.00	0.00	0.23

Source: Stephens, M., Johnson, D., Soward, J., and Ammenheuser, J., *Guide for the Design of Semiconductor Equipment to Meet Voltage Sag Immunity Standards*, International SEMATECH, 1999. Technology Transfer #99063760B-TR, available at http://www.sematech.org/public/docubase/document/3760btr.pdf.

The site data is not normally distributed. The site indices are nonnegative, and the distribution skews upward; therefore, we need another distribution, the log-normal, the Gamma, or the Weibull. Figure 2.12 includes fits to log-normal distributions. The median (M) of the log-normal distribution equals the mean of the natural log of the values (x_i) raised to e: $M = e^{\text{mean}(\ln(x_i))}$. The log standard deviation is $\beta = \text{sd}[\ln(x_i)]$.

2.5.2 Transmission-Level Power Quality

Large industrial customers, utility's prize customers, are primarily fed with transmission-level service and expect high-quality power. Several semiconductor manufacturing sites provided a basis for developing the SEMI F47 standard for semiconductor tools (Stephens et al., 1999). These sites were primarily served from transmission lines; not all were direct transmission services, but distribution exposure was minimal. While not as extensive as the DPQ study, the monitoring provides good data on the number of events that are primarily from the transmission exposure. Table 2.8 shows summary statistics from the SEMI dataset of 16 sites with 30 total monitor-years of data. Figure 2.13 compares distributions of SEMI data with the DPQ substation data. As expected, the semiconductor manufacturing sites experience fewer events compared to the typical DPQ site. This comparison provides some guidance on the portion of distribution events that are caused on the transmission system. Use caution though since these are two independent data sets.

2.6 Correlations of Sags and Momentaries

Figure 2.14 shows the number of momentary interruptions at a site plotted against the number of voltage sags. We see that sites with high numbers of

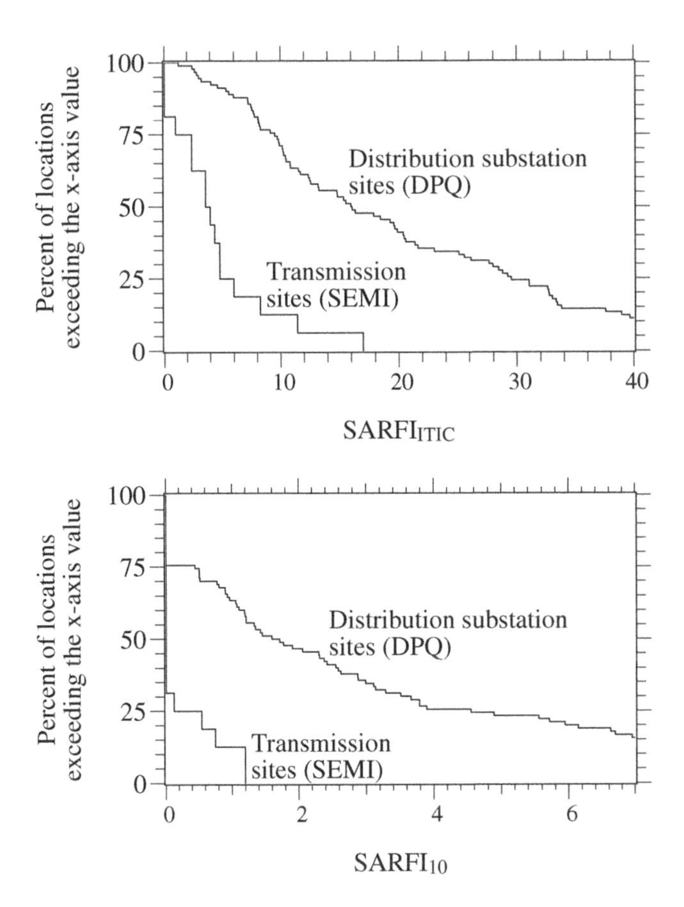

FIGURE 2.13
Comparison of the 16 SEMI sites with the DPQ substation sites.

momentary interruptions probably also have high numbers of voltage sags. Sites with low numbers of momentary interruptions may have high or low numbers of voltage sags. The correlation coefficient between sags and momentaries for the DPQ sites is 44.8%.

Correlations between deep voltage sags and shallow sags are more pronounced. $SARFI_{90}$ and $SARFI_{50}$ have a 56.9% correlation coefficient. If we break the sites down by load density, the correlation coefficients improve to 90, 84, and 74% for urban, suburban, and rural sites.

2.7 Factors That Influence Sag and Momentary Rates

Power system faults cause voltage sags and momentary interruptions. The frequency of faults depends on many factors including weather, maintenance, and age of equipment. The protection schemes and location of circuit

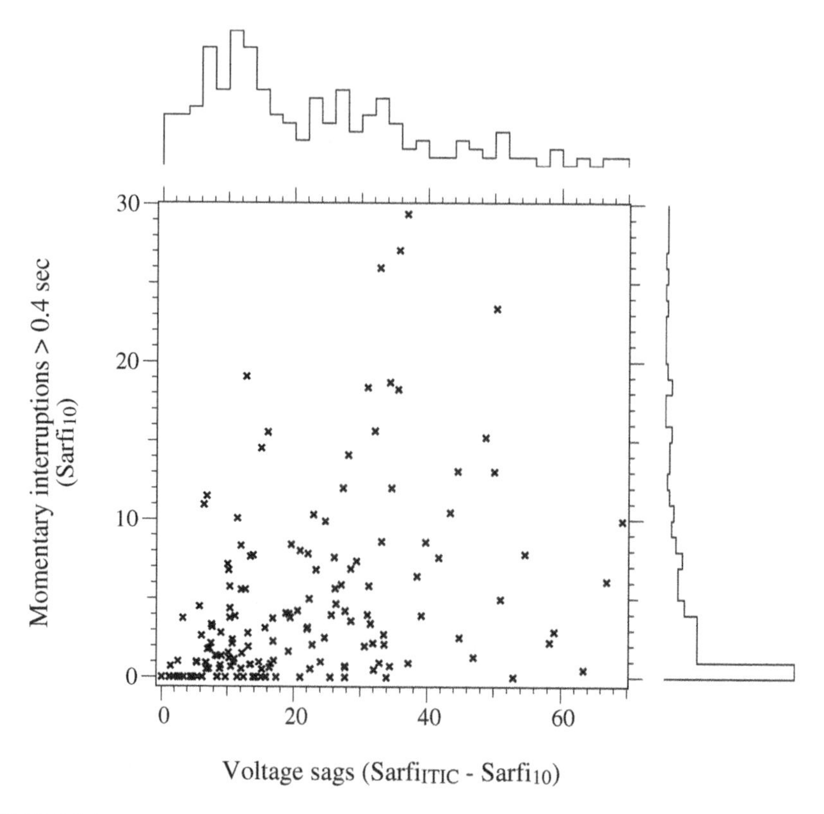

FIGURE 2.14
Relationship between voltage sags and momentary interruptions (greater than 0.4 sec). Each point gives the average voltage sags and momentary interruptions at a site (n = 158).

interrupters determine whether a fault causes a voltage sag or an interruption, and the protection system determines the event duration. The following sections describe work using EPRI's DPQ data to investigate what factors influence sags and momentaries (Short et al., 2002).

2.7.1 Location

Three monitors were used on each circuit in the DPQ study. One was always at the substation, and two were on the feeder, named "feeder middle" and "feeder end." The feeder sites were randomly picked on the circuits, so the naming is somewhat misleading; "feeder end" does not mean the most distant point from the substation (it just means the most distant of the two monitors randomly placed on the circuit). Since one third of the monitors are at the substation, the set is biased to "near-substation" customers since most customers are not located near the substation. Although there is some difference between measurement locations, it turns out that it is not drastic.

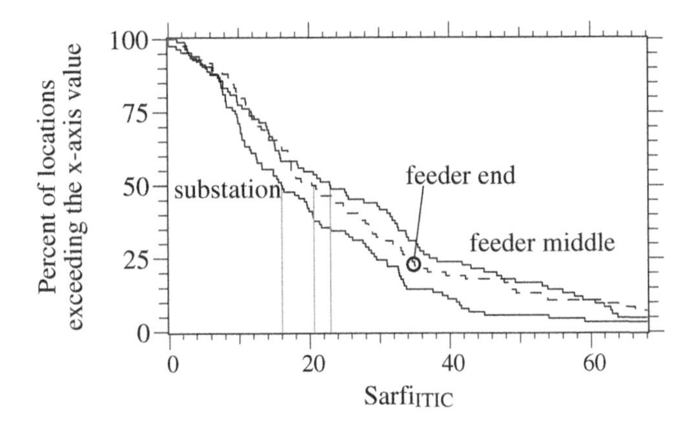

FIGURE 2.15
Comparison of feeder sites and substation sites in the DPQ data for $SARFI_{ITIC}$.

There is surprisingly little difference between the distributions of monitoring locations (see Figure 2.15 for $SARFI_{ITIC}$).

Figure 2.16 shows a more specific comparison of the substation's performance plotted against its two feeder sites. As expected, most feeder sites have more sags than their substation site, especially rural sites. A significant number of feeder sites were better than the substation. Measurement anomalies could produce this (the substation recorder is down for part of a bad storm season), or it could be real (downstream regulation devices keep the nominal voltage higher or the connected load "pushes" back on the source

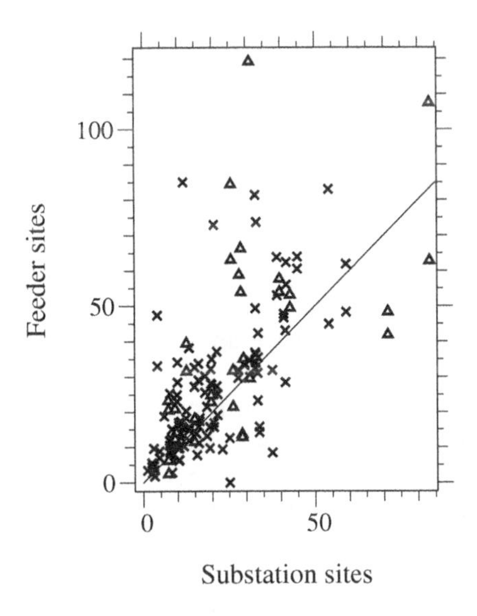

FIGURE 2.16
$SARFI_{ITIC}$ at substation sites plotted against $SARFI_{ITIC}$ at that substation's feeder sites (triangles indicate rural sites).

TABLE 2.9

Statistics for Momentary Interruptions Longer than 0.4 sec

	P(75%)	Median P(50%)	P(25%)
Rural	2.37	8.56	18.31
Suburban	0.23	2.39	6.71
Urban	0.00	1.37	2.82

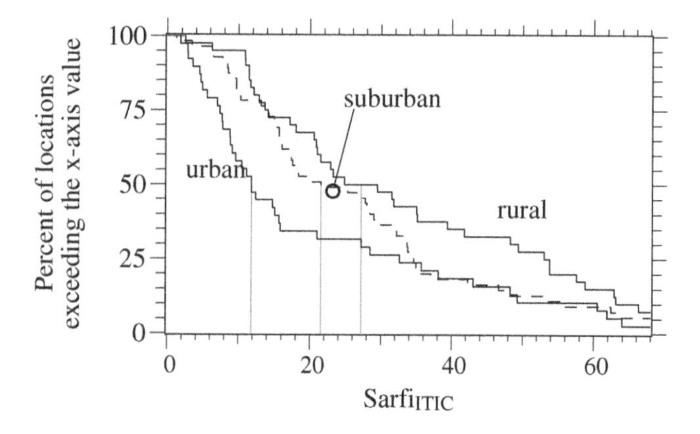

FIGURE 2.17
Comparison of urban, suburban, and rural sites for $SARFI_{ITIC}$ (feeder sites only).

impedance during bus faults). For most of our analysis, we excluded substation sites, thinking that the feeder sites better represent a random feeder location where customers are fed.

2.7.2 Load Density

Rural sites have more voltage sags and momentary interruptions (see Table 2.9 and Figure 2.17). This is not surprising given the extra lengths of line needed to serve load in low-density areas. Interruptions showed the most dramatic difference.

Why do urban sites not have even more profoundly lower voltage sag rates than suburban and rural sites? After all, urban sites are shorter and mostly underground with fewer faults per mile. The main answer is that urban sites have many more feeders off of a bus. In addition, even though urban circuits are shorter, most of the exposure is close to the substation. So, while many of the faults on rural and suburban circuits are too far away to pull down the substation voltage, almost every fault on an urban circuit causes a significant voltage sag for all customers off of that substation bus.

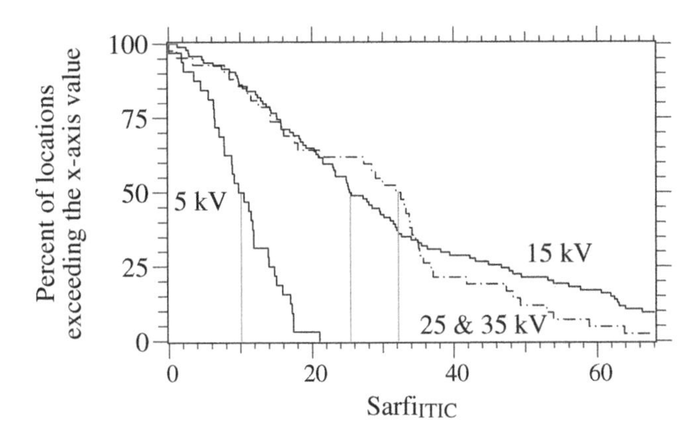

FIGURE 2.18
Comparison of feeder sites by voltage class in the DPQ data.

2.7.3 Voltage Class

Figure 2.18 shows that 5-kV systems have much lower numbers of voltage sags and interruptions. Lower voltage systems have less feeder exposure, and higher line impedance relative to the station transformer. Fault rates are often lower on 5-kV systems. Somewhat surprisingly, the 25 and 35-kV systems were not worse than the 15-kV systems.

2.7.4 Comparison and Ranking of Factors

We analyzed data available on the DPQ site characteristics to determine what parameters most affected power quality events. Figure 2.19 shows the variations of $SARFI_{ITIC}$ with site characteristics.

The three most significant predictors of excursions below the lower ITI curve are:

1. *Circuit exposure* — The total exposure on the circuit including three-phase and single-phase portions is a good predictor of voltage sags. Any fault on the circuit sags the voltage.
2. *Lightning* — Lightning causes many faults on distribution systems, and lightning strongly correlates with voltage sags (based on the 10-year average, 1988–98, from the U.S. National Lightning Detection Network). In addition, lightning predicts weather patterns — areas with high lightning tend to have more storms and more wind and tree-related faults.
3. *Transformer impedance and number of feeders* — The $n_f \cdot kV^2/MVA_{xfrmr}$ term in Figure 2.19 contains the number of feeders off of the transformer bus along with an estimate of the transformer impedance. The transformer impedance is $Z_{\%}kV^2/MVA$; but since the per-unit

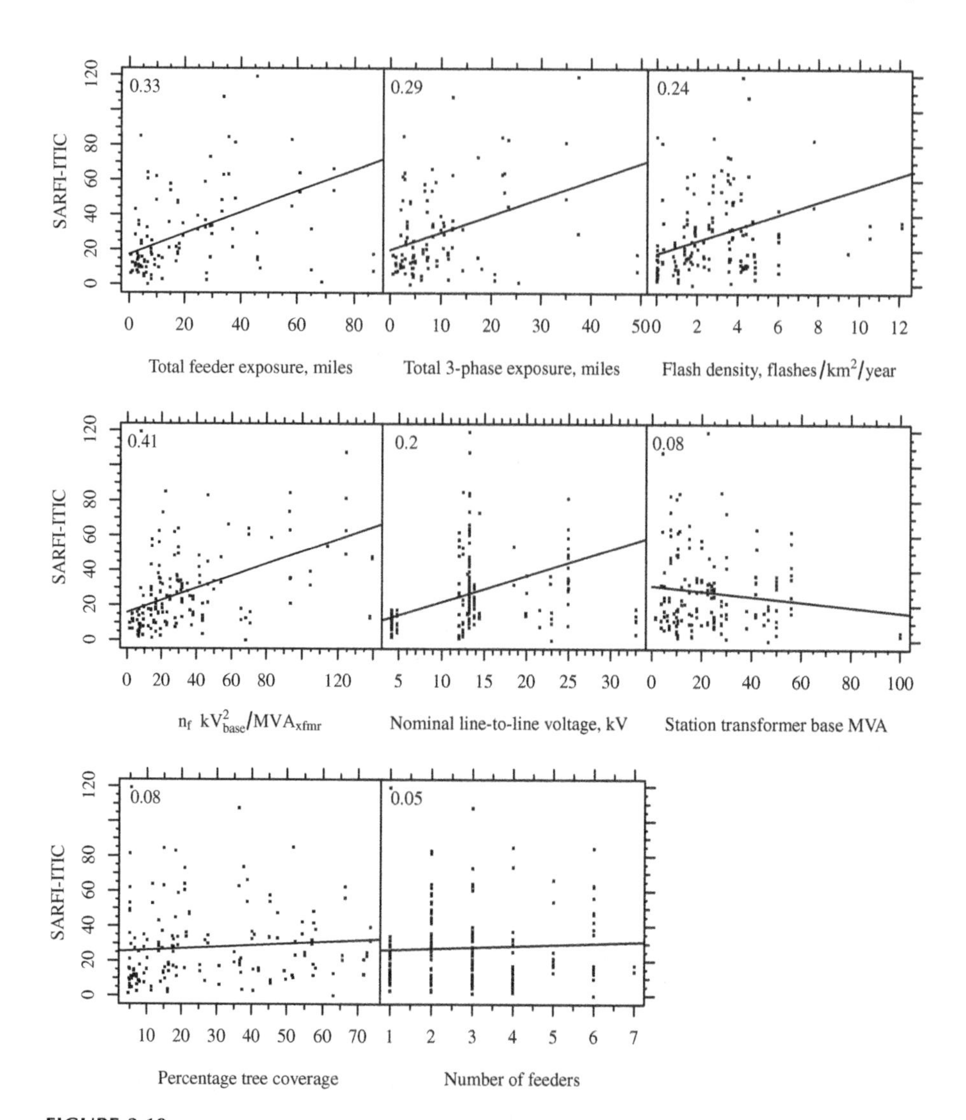

FIGURE 2.19
Variations in the number of excursions below the lower ITI curve (which are mainly voltage sags) vs. various site parameters. The correlation coefficients (r) are given in the upper-left corner of each plot.

impedance of station transformers is roughly constant (7 to 10%), we use kV^2/MVA.

This last term requires a bit more explanation. The number of bus sags is directly proportional to n_f, the number of feeders off the bus and to Z_S, the source impedance (a lower station transformer impedance — a bigger transformer or lower percent impedance — improves voltage sags at the station bus). We approximate these two terms as $n_f \cdot kV^2/MVA_{xfrmr}$

Other variables have much less impact on the number of voltage sags than the three main parameters given.

2.8 Prediction of Quality Indicators Based on Site Characteristics

We derive a formula for predicting the number of events for a quality indicator based on a few of the characteristics of the site. If no measurement or historical data is available, this is useful in estimating the utility-side quality.

Regression techniques are commonly used to find a model prediction formula. A generalized linear model is a least-squares fit to an equation of the following form:

$$y = a_1 x_1 + a_2 x_2 + \cdots + a_n x_n + \varepsilon$$

The x's are site characteristics (such as base voltage or lightning flash density), and the a's are coefficients fitted to the model. The generalized linear model is somewhat different from a standard linear model; we used a generalization where the distribution of the error ε is assumed to be a gamma distribution rather than a normal distribution in a strictly linear model. A gamma distribution skews to the right, like the log-normal distribution.

A model for estimating $\text{SARFI}_{\text{ITIC}}$ is

$$N_{ITIC} = 4.74 + 0.293l + 2.47N_g + 0.192\frac{n_f \cdot kV^2}{MVA_{xfmr}}$$

$$+8.2 \text{ if moderate to heavy tree coverage}$$

where

N_{ITIC} = predicted annual number of events which fall under the lower ITI curve

l = total exposure (including three-phase and single-phase portions) on the circuit, mi (multiply kilometers by 1.609)

N_g = lightning ground flash density, flashes/km^2/year

kV = base line-to-line voltage, kV

n_f = total number of feeders off of the substation bus

MVA_{xfmr} = station transformer base rating (open-air rating), MVA

If any of the circuit characteristics are unknown, we could use the following medians from the DPQ data:

where

$$l = 14.5 \text{ mi } (23.4 \text{ km})$$
$$N_g = 2.57 \text{ flashes/km}^2\text{/year}$$
$$\frac{n_f \cdot kV^2}{MVA_{xfmr}} = 25$$

All three variable terms in the linear regression are significant to at least 99% (there is less than a 1% chance that the terms of the model do not influence the prediction). The tree coverage term is less certain; there is a 9% chance that the term is not significant. We based the tree coverage term on the University of Maryland's Global Land Cover Facility data from the Advanced Very High Resolution Radiometer (AVHRR). Half of the DPQ sites had more than 19% of the land area covered by trees, which we defined as "moderate to heavy tree cover."

How good is the model? It is decent given all the factors that affect sags and momentary interruptions and inherent variability. Given the variability of power quality events, it is surprising that the model is this good: 34% of the values are within 25% of the prediction, and 60% of the values are within 50% of the prediction. See Figure 2.20 for the prediction scatter.

For an example 12.47-kV case with three feeders, a 25-MVA transformer, a flash density of 4 flashes/km²/year, moderate tree coverage, and a total exposure of 32 km, the model predicts 29.8 events per year. For this case, the data shows a prediction interval with a 50% confidence level of between 15.6 and 34.2 events per year (the 90% confidence prediction interval is between 0 and 68.3). The data is so dispersed that the model is not good enough to use for precision estimates (such as in a contract for premium power).

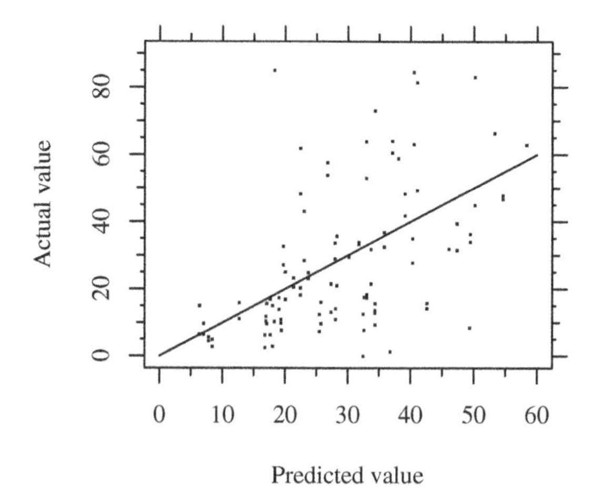

FIGURE 2.20
Actual values vs. predicted values for the model predicting the annual average number of events below the lower ITI curve.

The site characteristics most affecting sags but not included in this model (because no information was available) are (1) subtransmission exposure and characteristics and (2) percentage of the circuit that was underground.

A reasonable model for predicting momentary interruptions is

$$N_{10} = \begin{pmatrix} 5.52 \text{ if Rural} \\ 0.29 \text{ if Suburban} \\ -1.61 \text{ if Urban} \end{pmatrix} + 0.116 l_3 + 0.27 N_g + 1.24 \frac{n_f \cdot kV}{MVA_{xfmr}}$$

where

N_{10} = the predicted annual number of events with voltage less than 10% of nominal for more than 0.4 sec

l_3 = the three-phase circuit exposure, mi

The parameters differ somewhat from $SARFI_{ITIC}$ predictors. Two of the strongest indicators of momentary interruptions are load density and three-phase circuit exposure. Other significant parameters are the lightning activity and a term with voltage, number of feeders, and transformer MVA. The model is not as good as the ITI model, but all parameters have more than a 95% probability of affecting the result. The site characteristic most affecting momentaries that is not included in the model for lack of information is whether fuse saving is used.

2.9 Equipment Sensitivities

2.9.1 Computers and Electronic Power Supplies

Computers and other equipment with electronic power supplies are the most widely found equipment that is sensitive to power quality disturbances. The power supply is typically a switched mode power supply as shown in Figure 2.21. Computers have a wide range of sensitivities. The ride-through capability for interruptions of several computers is summarized in Figure 2.22. Many of the computers had ride through of more than 0.1 sec (0.28 sec was the best of this set of studies), and some could not even ride through a 0.01-sec interruption.

The ride-through capability of computers is close to rectangular. Two points describe the characteristic on a volt-time curve: the interruption ride through time and the steady-state ride-through point. There is usually a steep transition between the interruption ride-through point and the steady-state ride-through point. Other characteristics of computer ride through are:

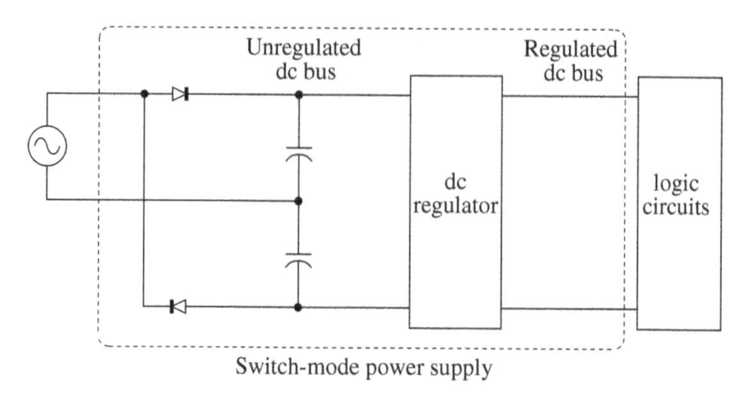

FIGURE 2.21
Switch-mode power supply used in most computers.

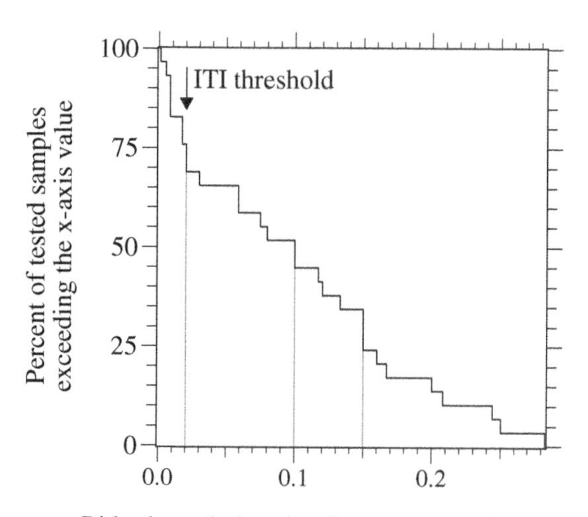

FIGURE 2.22
Capability of computers to ride through an interruption (n = 27). (Data from [Bowes, 1990; Chong, 2000; Courtois, 2001; Courtois and Deslauriers, 1997; EPRI PEAC Brief No. 7, 1992].)

- There is little difference between the performance when the computer is processing or accessing disk and when the computer is idle.
- The point on the waveform when the disruption occurs does not matter.

Figure 2.23 shows volt-time sensitivities of computers from several studies. The most sensitive units, those that violate the ITI curve, were made and tested before the ITI curve was created. That is not much of an excuse as the most sensitive computers also violated the CBEMA curve which was available to the manufacturers at that time (remember, there is no standard that requires testing computers to meet the ITI/CBEMA curve).

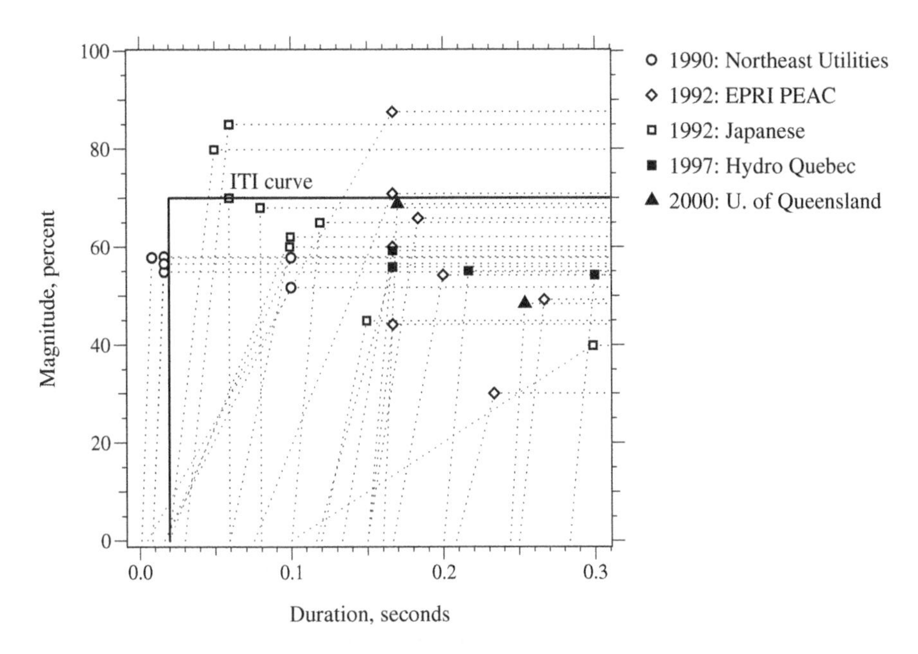

FIGURE 2.23
Volt-time characteristics of several computers tested in different studies. (Data from [Bowes, 1990; Chong, 2000; Courtois, 2001; Courtois and Deslauriers, 1997; EPRI PEAC Brief No. 7, 1992; Sekine et al., 1992]. Figure copyright © 2002. Electric Power Research Institute. 1007281. *Analysis of Extremely Reliable Power Delivery Systems.* Reprinted with permission.)

An important factor regarding the ride-through capability of computers is that it varies significantly depending on the voltage just before the interruption. The energy storage in a switch-mode power supply is from the front-end rectifier capacitors. The energy stored in a capacitor is $1/2CV^2$. Power supplies typically have two 470-μF capacitors in series, and the voltage across the two capacitors in series is $V_p = 2\sqrt{2} \cdot 120 = 339.4 \text{ V}$. We can estimate the ride-through capability of a computer as:

$$t = \frac{C(V_p^2 + V_d^2)}{4 \times 10^6 P}$$

where

t = ride-through duration for an interruption, sec
P = load on the computer, W
C = capacitance on one half of the bridge rectifier, μF (470 is common)
V_p = peak of the ac voltage, V (339.4 V for 120 V nominal)
V_d = voltage on the unregulated dc bus where the computer will drop out (use half of V_p if unknown or 0 for the maximum ride through)

Since the energy is a function of V^2, a voltage of 90% of nominal means the capacitor stores only 81% of the energy that it would at nominal voltage.

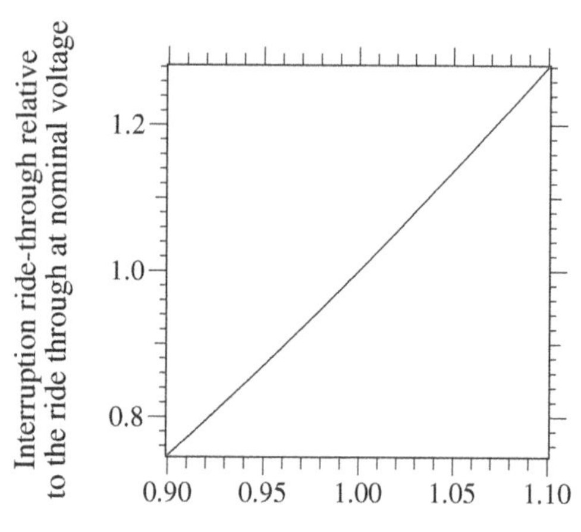

FIGURE 2.24
Change in the ride-through capability of computers vs. the voltage prior to the interruption.

Even worse, the computer drops out before all of the energy in the capacitor is used. Figure 2.24 shows the relative ride through as a function of the voltage prior to the interruption, assuming the computer drops out when the unregulated dc bus voltage reaches half of nominal.

The pre-disturbance voltage affects ride through for any device that has capacitance for energy storage, including most computer power supplies, programmable-logic controllers, digital clocks, and adjustable-speed drives. So, either on the utility side or the customer's side, raise voltages to inexpensively increase ride through of devices.

Even more ride-through capability is possible with computers. EPRI PEAC has done tests of a computer power supply modified with extra ride-through capability (EPRI PEAC Brief No. 12, 1993). The enhanced supply, developed by the New England Electric Company, had an extra 4500 μF of capacitors installed in parallel with the existing capacitors, which increased the ride through from 0.175 sec to 1.8 sec. In the near future, ultracapacitors may supply an even more economical ride-through enhancement.

Intelligent power management might also increase ride through. Laptops and most desktops have sophisticated ways of managing power to conserve energy. We could apply similar techniques to short-duration power interruptions. The processor, disk, and other power-hungry equipment could be "quick suspended" during a power interruption to increase the normal 0.05 to 0.2-sec ride-through capability. Just suspending the processor (30–50 W typically in fast, hot chips) would extend the ride through considerably. This enhancement requires very little extra hardware; a sensor to measure the incoming ac power or the unregulated dc bus voltage would be needed — no microprocessor-level changes are required.

Industrial dc power supplies share the same characteristics as the computer power supply. Heavily loaded power supplies are more susceptible to voltage sags and interruptions. Use a supply rated at twice the load on the supply to increase ride through.

A power supply with a universal input operates over a wide range of voltages (85 to 264 V typically), but the ride-through capability changes dramatically with operating voltage. Operation as close to the upper end as possible improves ride through. For this reason, prefer a line-to-line connection (208 V) over a line-to-ground connection (120 V). The low-voltage limit of 85 V is 71% of nominal at 120 V, but we obtain much better ride through when applied at 208 V (now the lower limit is 41% of nominal). The difference in $1/2CV^2$ is dramatic in the two cases. By the same token, if the power supply has alternate settings, use the setting that positions the actual voltage near the high end of the range. Consider a power supply with a 95 to 250-V range designed for Japanese and European loads and a 110 to 270-V range designed for America. The range with an upper limit of 250 V for a 208-V line-to-line connection results in the maximum ride through (McEachern, 2001).

Another option for some industrial supplies and large computer systems is a three-phase power supply instead of a single-phase supply. A three-phase supply is less sensitive to voltage sags. Single-phase sags only slightly depress the dc bus voltage of a three-phase rectifier because the remaining two phases can fully supply the load. Even a two-phase sag is significantly less severe than a three-phase sag.

Linear power supplies have much less ride-through capability than switch-mode power supplies (switch-mode supplies may have 100 times the capacitance). Fortunately, most power supplies are switch-mode supplies (primarily because they are lighter, more efficient, and cost less).

2.9.2 Industrial Processes and Equipment

A variety of industrial equipment is sensitive to voltage sags. Some of the main sensitive equipment used in industrial facilities are

- Programmable logic controllers (PLCs)
- Adjustable-speed drives (ASDs), also called variable-speed drives (VFDs)
- Contactors
- Relays
- Control equipment

Depending on the process and load, any number of devices can be the weak link. Table 2.10 shows the breakdown of weak links for semiconductor tools serving the semiconductor manufacturing industry.

TABLE 2.10

Breakdown of Semiconductor-Tool Voltage Sag Sensitivities (n = 33)

Weak Link	Overall Percentage
Emergency off (EMO) circuit: pilot relay (33%), main contactor (14%)	47%
dc power supplies: PC (7%), controller (7%), I/O (5%)	19%
3-phase power supplies: magnetron (5%), rf (5%), ion (2%)	12%
Vacuum pumps	12%
Turbo pumps	7%
ac adjustable-speed drives	2%

Source: Stephens, M., Johnson, D., Soward, J., and Ammenheuser, J., *Guide for the Design of Semiconductor Equipment to Meet Voltage Sag Immunity Standards*, International SEMATECH, 1999. Technology Transfer #99063760B-TR, available at http://www.sematech.org/public/docubase/document/3760btr.pdf.

2.9.2.1 *Relays and Contactors*

Contactors are electromechanical switches used for a variety of power and control applications. A contactor uses a solenoid to engage when an appropriate voltage is applied. More voltage is required to close the contactor than is required to keep it closed.

Relays and contactors can drop out very quickly. Figure 2.25 shows the ride-through duration for an interruption for several relays and contactors, and Figure 2.26 shows the dropout levels for voltage sags. The devices are somewhat dependent on the point on the wave where the voltage sag starts. Ride through is longest for sags starting at the voltage zero crossing; but unfortunately, faults tend to occur when the voltage is near its peak. The fast dropout of contactors limits some of the utility-side solution options — faster relaying, smaller fuses, or 1.5-cycle transfer switches may provide good improvement to computers but offer little help for many relays and contactors. Because they trip very quickly, voltage mainly dominates, not the duration.

The volt-time capability of relays and contactors approximates a rectangular shape. Contactors can have the unusual property that the ride-through capability improves at lower voltages. An example volt-time ride-through characteristic is shown in Figure 2.27. The reason for this property relates to the fact that current, and not voltage, holds a contactor in. A contactor contains shading rings, which are analogous to damper windings in a rotating machine. A shading ring is a shorted winding around the magnetic core. In response to a voltage transient, the shading ring produces a back emf that opposes the transient. A larger transient (deeper sag) creates more current that holds the contactor in (Collins Jr. and Bridgwood, 1997).

A larger relay generally has more ride through; a contactor usually has more ride through than a relay. Some of the most sensitive relays are small industrial relays with clear plastic cases referred to as *ice-cube* relays.

Several options are available to help hold in contactors and relays (St. Pierre, 1999):

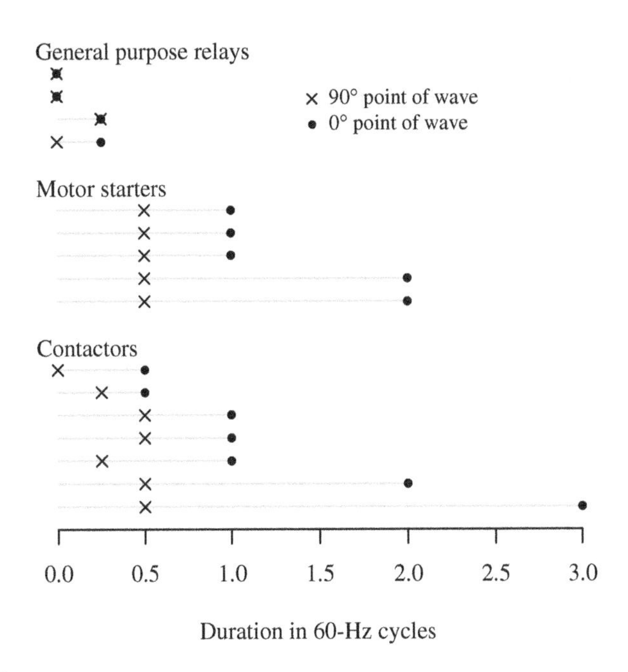

FIGURE 2.25
Ride-through duration for an interruption to several relays and contactors. (Data from [EPRI PEAC Brief No. 44, 1998].)

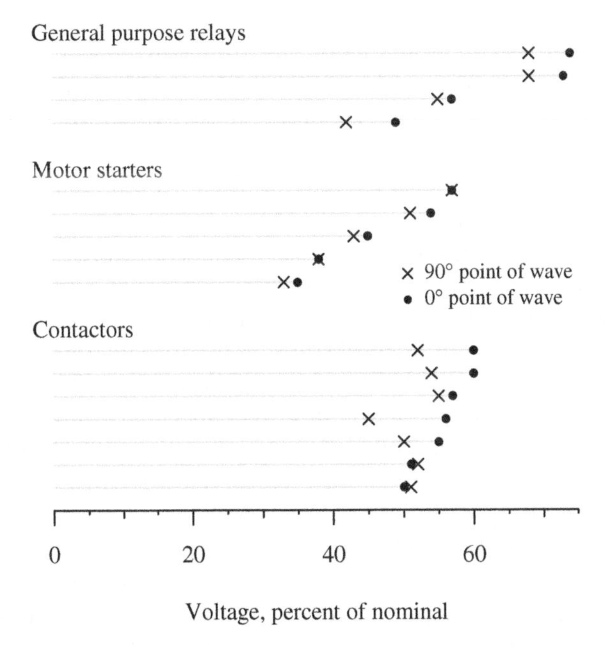

FIGURE 2.26
Voltage magnitude for dropout of several relays and contactors for a five-cycle voltage sag. (Data from [EPRI PEAC Brief No. 44, 1998].)

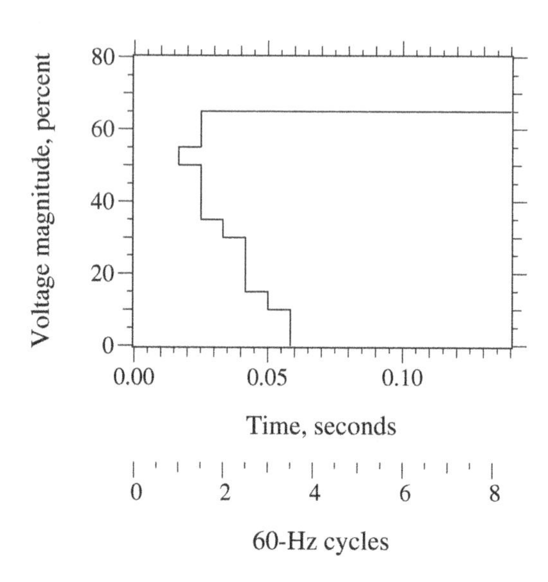

FIGURE 2.27
Ride-through capability of a NEMA size 1 contactor rated 0 to 50 hp at 480 V. (Data from [EPRI PEAC Brief No. 10, 1993].)

- *Coil hold in* — Use a coil hold-in device. Coil hold-in devices supply current to keep a relay or contactor coil held in during a voltage sag to about 25%.
- *dc* — Rectify the ac voltage and use a dc contactor. A dc contactor generally has a longer ride through than an ac contactor. A capacitor added in parallel with the contactor coil can extend the ride-through time.
- *Time-delay relay* — Add a time-delay relay to the control circuit in parallel with the contactor. If the contactor drops out because of a voltage sag, the time-delay relay keeps the control circuit energized (but the motor still drops out). If the sag finishes before the time-delay setting has elapsed, the contactor pulls back in, and the motor reconnects. This is not the best solution because the motor is disconnected and reconnected. The reconnection draws inrush current, which can itself cause local disruptions, especially if multiple motors are energized together. The high current may trip facility relays and possibly damage motors. Additionally, the disconnection of many motors within a facility removes the voltage support provided by the motors feeding back into the utility system. Furthermore, when the motors reconnect, the inrush creates a voltage sag.
- *Power conditioner* — Apply a constant voltage transformer or other power conditioning device such as a dip-proofing inverter to the control circuit. This provides enough ride through for all but the deepest voltage sags.

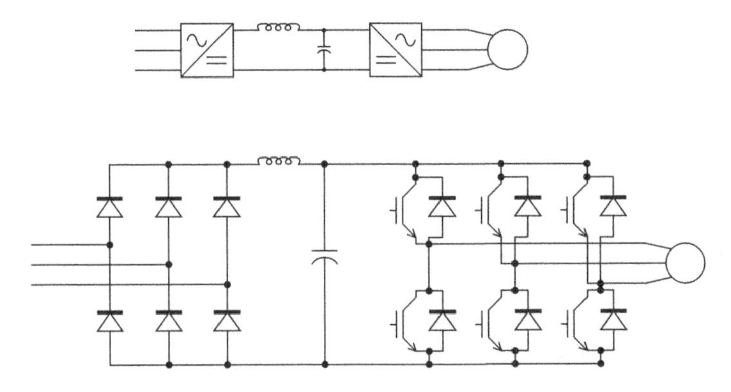

FIGURE 2.28
Common adjustable-speed drive topology.

2.9.2.2 Adjustable-Speed Drives

Adjustable-speed drives (ASDs) are very common industrial tools used to perform a variety of tasks. Figure 2.28 shows the most common drive topology: the three-phase incoming supply is rectified to dc, and a pulse-width modulation (PWM) inverter converts the dc to a variable frequency three-phase ac voltage that drives an induction motor at variable speeds.

The capacitor on the dc bus provides some energy storage but not much. Adjustable-speed drives are sensitive to voltage sags and almost always drop out for momentary interruptions. A common cause for shutdown is that the dc bus drops too low, and the drive shuts down on an undervoltage trip (normally 70 to 85% of the nominal dc bus voltage). Even if a drive does not actually shut down, the motor may stall and not be able to start without removing the mechanical load. External factors can also cause shutdown: if the drive is wired through a contactor and the contactor trips, the drive shuts down (and the contactor may be more sensitive than the drive) (EPRI PEAC commentary #3, 1998). Also, if programmable logic controllers provide stop, start, or other signals, loss of the controller can shut the drive down. Following a voltage sag, the drive can draw large inrush, as much as three or four times normal current. This can blow fuses or damage the input diodes and trip the drive.

Drives have significant variation. Of small drives tested by EPRI PEAC (Brief No. 9, 1993) at full load, two could ride through five-cycle three-phase sags down to zero volts; another two tripped at about 80% of rating. At half load, three drives survived a five-cycle sag down to zero volts, and the other tripped at 70% of rating. Some of the most sensitive drives are:

- *Older drives* — Many older drives power their electronics from the ac system, which makes the controller more sensitive. Modern drives power the controls from the dc bus.
- *Higher horsepower drives* — The front-end circuit normally uses thyristors instead of diodes in a current-source topology. To prevent

commutation failure, the dc undervoltage relay is set more sensitively, often 85 to 90%.

- *dc drives* — A thyristor-bridge feeds directly into the dc motor armature. dc drives have no built-in energy, and sags can disturb the timing circuitry for firing the thyristors. Regenerative converters may be very sensitive, especially during regeneration (reverse power flow); a voltage sag can prevent a thyristor from shutting off, which puts a short on the system that will blow a fuse.

Adjustable-speed drives are less sensitive to single-phase sags than three-phase sags because all three phases are rectified (Mansoor et al., 1997). A three-phase sag sags the dc bus voltage down in similar proportions. With a single-phase sag, the two "unsagged" phases can support the drive's dc bus. Whether the drive trips or not depends on how heavily it is loaded and how the undervoltage detection circuitry works. Figure 2.29 shows an example ride-through capability for a 60-kW drive.

Configuration adjustments can sometimes improve ride through; reducing the undervoltage trip setting significantly improves ride through. Increasing the overcurrent trip setting and setting appropriate restart parameters can also help. Also, some models have firmware upgrades providing additional ride through. Drives with a flying restart feature (the drive can restart while the motor is spinning) are better for critical loads. A drive with a synchronous flying restart following a sag to 50% voltage for five cycles allowed only a 5% decrease in motor speed and was fully restored in 1/2 sec (a nonsynchronous flying restart is not nearly as good) (EPRI PEAC Brief No. 30, 1995).

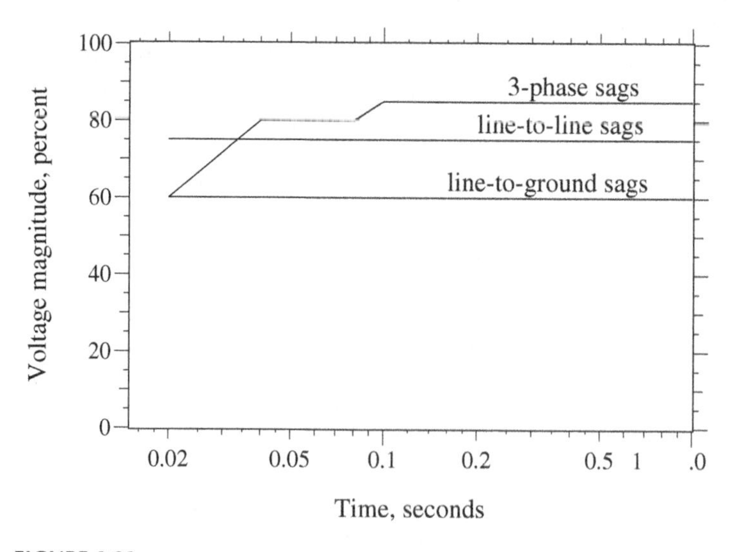

FIGURE 2.29
Ride-through capability of a 60-kW adjustable-speed drive under different voltage sag conditions. (Data from [Abrahams et al., 1999].)

2.9.2.3 Programmable-Logic Controllers

The performance of programmable-logic controllers (PLCs) varies widely. A PLC is a hardened electronic controller used to control many types of industrial processes and equipment. PLCs have multiple input and output channels (I/O racks) used to measure and control equipment. A dc power supply powers the cpu and the I/O racks.

In most cases, the PLC power supply is the same as other computer power supplies, a switched-mode power supply usually capable of riding through an interruption of several cycles. The problem is that the power supply is often not the weakest link in the system. A power supply monitoring circuit that senses the input voltage may initiate a shutdown during a voltage sag. PLCs are more sensitive as a result.

Another important concern is that sags or interruptions can not only cause a shutdown of the PLC, a sag can produce faulty outputs on some PLCs. Faulty outputs can cause more havoc with some processes than if the PLC actually shut down. Figure 2.30 shows the sensitivity of several programmable-logic controllers.

2.9.3 Residential Equipment

The digital clock has been quoted as being "the world's best-selling power quality recorder." The "blinking clocks" are a nuisance for customers and generate many phone calls for utilities. That said, clocks have a wide range of sensitivities, and many actually have very good ride-through capability.

Figure 2.31 shows ride-through capabilities for several digital clocks tested by EPRI PEAC and Hydro Quebec. There is a wide range of voltage sensitivity, but few of the digital clocks tested (25%) lose memory for a complete interruption that is less than 0.5 sec. The main consideration for distribution circuits is the dead time on the first reclose attempt. This is usually about 0.3 to 5 sec depending on the delay on the reclosing relay. An immediate reclose attempt has a dead time of 0.3 to 0.5 sec, making it a good option for reducing blinking clocks. Table 2.11 shows the ride through of different residential devices from a study by Northeast Utilities that also shows most devices have good ride through for events less than 0.5 sec.

Most residential devices have a fairly rectangular volt-time characteristic. Figure 2.32 shows the characteristic of several residential devices. Only longer-duration voltage sags affect most residential devices.

2.10 Solution Options

2.10.1 Utility Options for Momentary Interruptions

We can reduce momentary interruptions in several ways:

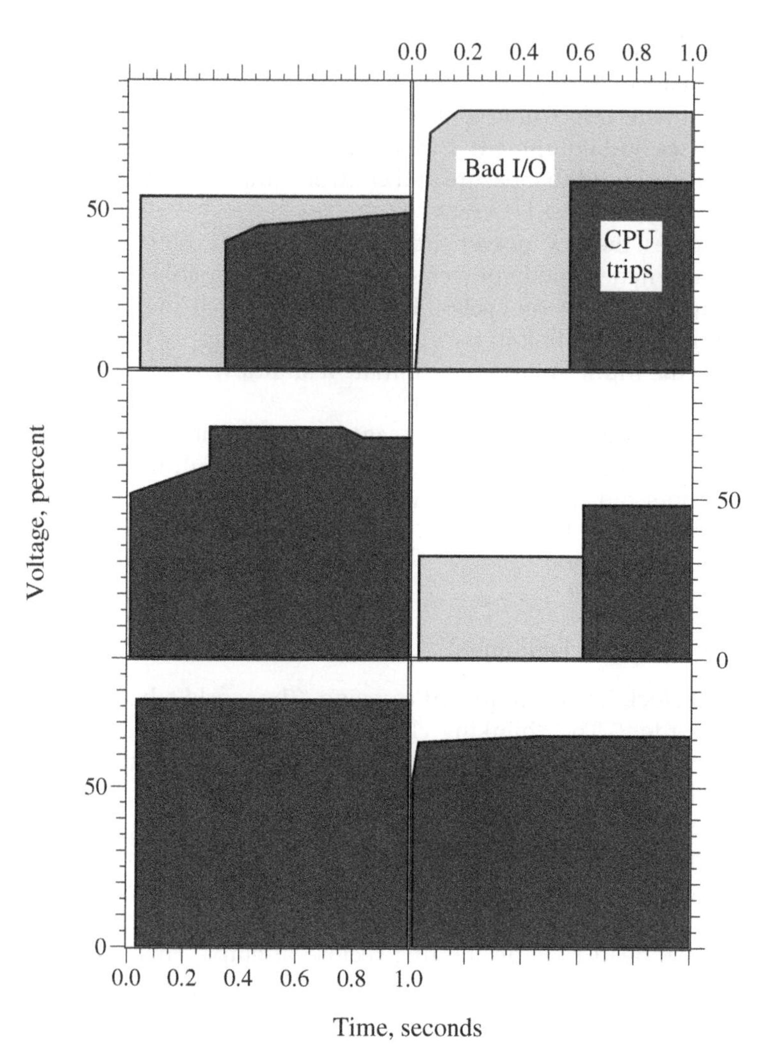

FIGURE 2.30
Sensitivity of six PLCs. (Adapted from [EPRI PEAC Brief No. 39, 1996].)

- Immediate reclose
- Use of fuse blowing
- Single-phase reclosers
- Extra downstream devices (fuses or reclosers)
- Sequence coordination with downstream devices
- Reduce faults

Reducing faults is a universally good approach to improving power quality. Other approaches target specific disturbances. For momentaries, the single

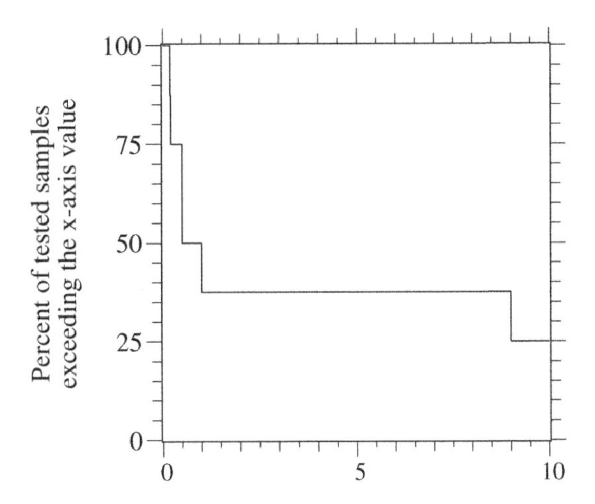

FIGURE 2.31
Capability of digital clocks to ride through an interruption (n = 8). (Data from [Courtois, 2001; Courtois and Deslauriers, 1997; EPRI PEAC Brief No. 17, 1994].)

TABLE 2.11
Percentage of Devices That Were Able to Successfully Ride Through a Momentary Interruption of the Given Duration

Device	0.5 sec	2 sec	16.7 sec
Digital clock	70	60	0
Microwave oven	60	0	0
VCR	50	37.5	0
Computer	0	0	0

Source: Bowes, K. B., "Effects of Power Line Disturbances on Electronic Products," *Power Quality Assurance Magazine*, vol. Premier V, pp. 296–310, 1990.

biggest improvement is to use fuse blowing instead of fuse saving. Using an immediate reclose, while not reducing the number of momentaries counted, reduces complaints from residential customers. Both of these changes for improving momentaries are relatively easy to implement.

Other methods of reducing momentaries involve better application of protection equipment. Extra protection devices that segment the circuit into smaller sections help improve momentary interruptions (and long-duration interruptions). Improving coordination between devices (including use of sequence coordination to improve coordination between reclosers) helps eliminate some unnecessary blinks. Single-phase reclosers instead of three-phase reclosers or breakers helps reduce the number of phases interrupted for single line-to-ground faults.

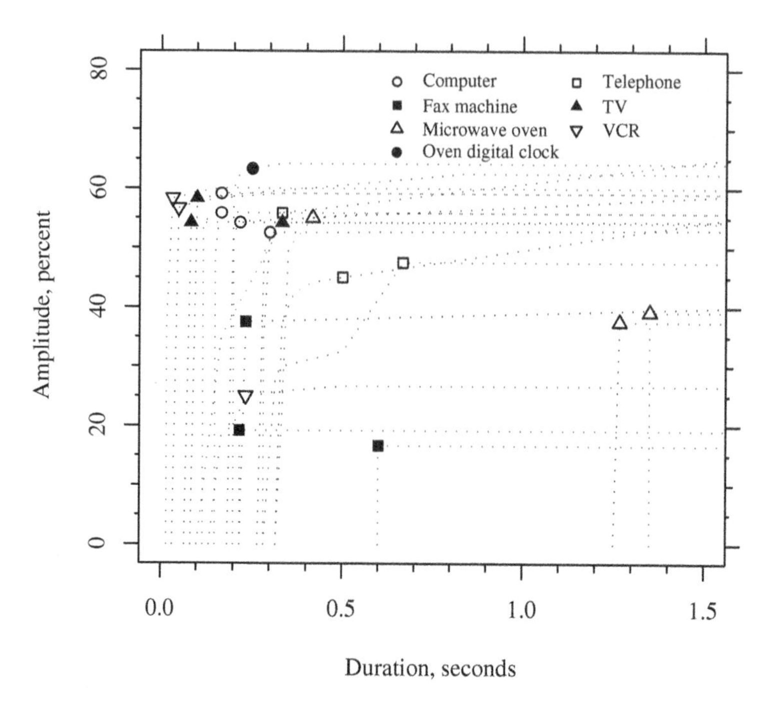

FIGURE 2.32
Ride-through capability of various residential devices. (Data from [Courtois, 2001; Courtois and Deslauriers, 1997].)

2.10.2 Utility Options for Voltage Sags

Utility-side options for reducing voltage sags are limited and are rarely done (at least solely for the purposes of voltage sags). Some of the strategies and equipment that help are:

- Use of fuse saving
- Current limiting fuses
- Smaller lateral fuses
- Faster breakers or reclosers
- Raise the nominal voltage
- Reduce faults

Faster relaying or faster interrupters — any changes in protection schemes that clear faults faster — help reduce the sag's duration. The next few sections address some of the options for reducing the impact of power-system faults on the voltage.

2.10.2.1 Raising the Nominal Voltage

Raising the nominal voltage helps the ride through of many types of equipment. Computers, adjustable-speed drives, and other equipment with capacitors benefit if the voltage is regulated near the upper end of the ANSI range A, or at least avoid the lower end. On the utility system, we use LTCs, regulators, and switched capacitors. This approach is the opposite of demand-side management programs which are designed to deliver low voltages in order to reduce peak demand or customer energy usage.

2.10.2.2 Line Reactors

Series line reactors provide electrical separation between feeders off the substation bus. Figure 2.33 shows the effect of line reactors on the station bus voltage for different configurations. Reactors have the added benefit of limiting the fault current. The reactors provide good protection against some

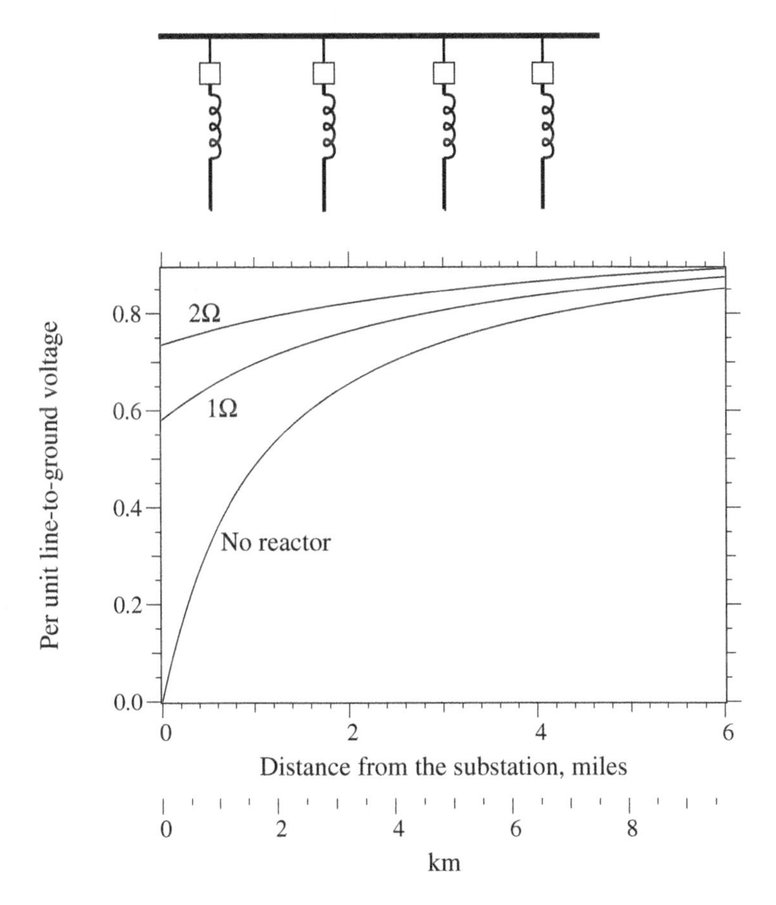

FIGURE 2.33
Substation bus voltage for a three-phase fault at the given distance for different line reactor configurations (for a 12.5-kV circuit with the same parameters as Figure 7.11).

voltage sags but do not help with interruptions or sags caused by transmission-level faults. On the faulted feeder, line reactors make the voltage sag worse, but much of the feeder may have very low voltage anyway (downstream of the fault), or the breaker may open and give all of the customers an interruption.

Utilities have used line reactors in this application, mainly at urban stations to reduce high fault currents. Line reactors do not have the best reputation; they are expensive, take up substation space, and increase voltage drop. The reactor must be designed to withstand the fault currents that it will regularly see.

2.10.2.3 Neutral Reactors

Phase reactors effectively isolate feeders and limit the voltage sag, so how about the reactor sometimes used between the substation transformer neutral and ground? The answer depends on the transformer connection serving the customer and the connection of the load. Neutral reactors generally make voltage sags worse for loads connected line-to-ground on the distribution circuit because:

- *Duration* — A neutral reactor lowers the fault current, so tap fuses and time overcurrent relays take slightly longer to operate, increasing the sag duration.
- *Magnitude* — A neutral reactor increases the zero-sequence impedance. The weaker ground source cannot hold up the station line-to-ground voltage as well for a downstream line-to-ground fault.

For line-to-line connected loads, the neutral reactor significantly improves voltages to end-use equipment, because of

- *Neutral shift* — The reactor adds impedance which shifts the neutral point. This raises the voltage on the unfaulted phases, but supports the line-to-line voltages. (A very large reactor in a high-impedance grounded system would have almost no drop in the line-to-line voltages during a single line-to-ground fault.)

With a single line-to-ground fault, the line-to-line voltage in per unit is (assuming that the circuit resistances are zero for simplification)

$$V = \frac{\sqrt{X_0^2 + X_0 X_1 + X_1^2}}{2X_1 + X_0}$$

The neutral reactor of X ohms adds $3jX$ ohms to the zero-sequence impedance and raises the X_0/X_1 ratio. Table 2.12 and Figure 2.34 show that a modest

TABLE 2.12

Line-to-Ground and Line-to-Line Voltages on the Low-Voltage Side of a Transformer with One Phase on the High-Voltage Side Sagged To Zero

Voltages	Primary Voltages			Voltages Downstream of a Delta – Wye Transformer		
No neutral reactor ($X_0/X_1 = 1$)						
Line-ground	0.00	1.00	1.00	0.58	0.58	1.00
Line-line	0.58	0.58	1.00	0.33	0.88	0.88
Neutral reactor that gives $X_0/X_1 = 3$						
Line-ground	0.00	1.25	1.25	0.72	0.72	1.00
Line-line	0.72	0.72	1.00	0.60	0.92	0.92

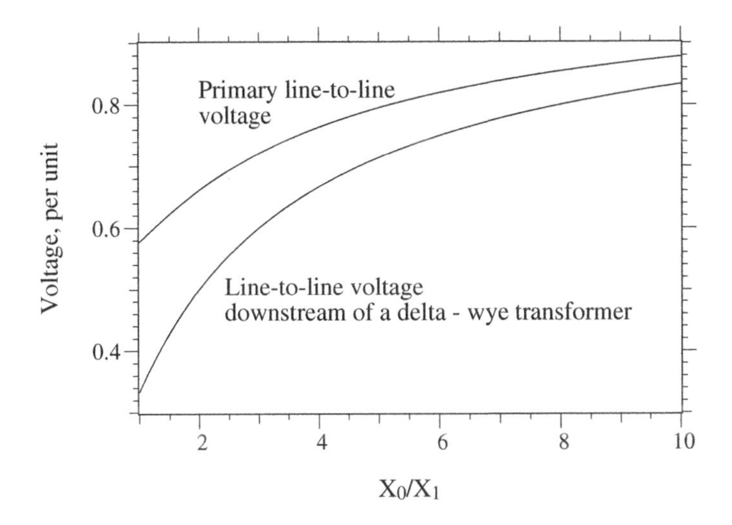

FIGURE 2.34
Impact of a neutral reactor (higher X_0) on voltage sags.

neutral reactor significantly helps line-to-line loads during line-to-ground faults. Also, with delta – wye distribution transformers, the line-to-ground secondary voltages see equivalent sags as the line-to-line voltage on the primary. The neutral reactor provides benefit only for line-to-ground faults and no help for three-phase or line-to-line faults. A disadvantage of the neutral reactor is that it increases the voltage rise (the swell) on the unfaulted phases.

2.10.2.4 Current-Limiting Fuses

Current-limiting fuses reduce the fault current and force an early zero crossing. In the process, the fuse reduces the severity of the magnitude and

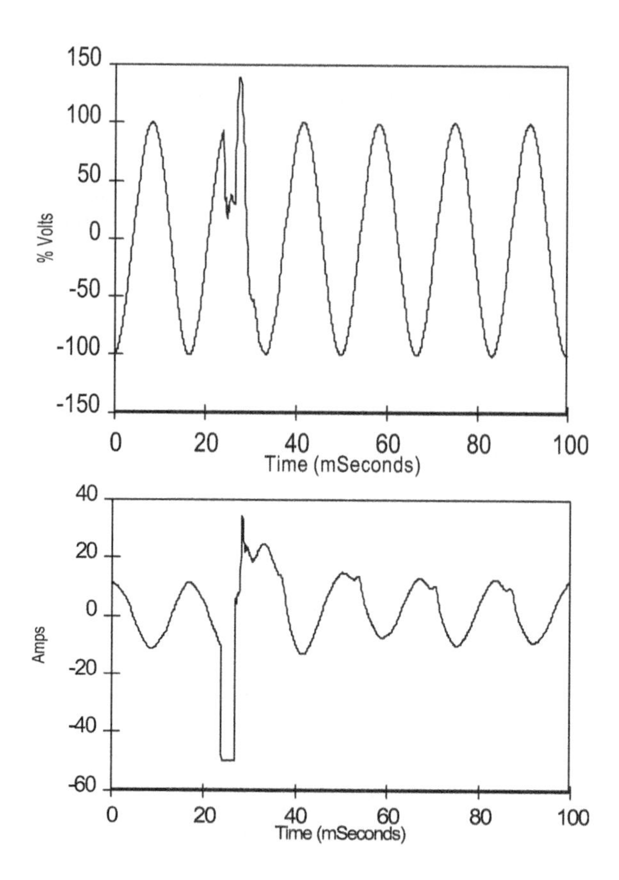

FIGURE 2.35
Example of a current-limiting fuse operation recorded in the EPRI DPQ study. (Copyright © 1996. Electric Power Research Institute. TR-106294-V3. *An Assessment of Distribution System Power Quality: Volume 3: Library of Distribution System Power Quality Monitoring Case Studies.* Reprinted with permission.)

reduces the duration. Figure 2.35 shows an example of a fault cleared by a current-limiting fuse. The duration of the sag is very short, and the depth is minimal. These types of results have been verified by other measurements and computer models (Kojovic and Hassler, 1997; Kojovic et al., 1998). Close to the substation is the best location for use, where they are most useful for reducing the magnitude and duration of voltage sags and are most appropriate for limiting damage due to high fault currents.

2.10.3 Utility Options with Nontraditional Equipment

2.10.3.1 Fast Transfer Switches

Medium-voltage static transfer switches are a utility-side option for providing improved power quality and protection from voltage sags and momentary interruptions. Thyristors are used to obtain a 1/2-cycle transfer between two sources. The switches can be configured as preferred/alternate or as a split bus. The normal configuration employs static transfer switches for sensitive customers in a primary selective scheme. After the switch has operated, it may or may not switch back to the original feeder.

We should also consider the impacts on the distribution system. Switching a large load between weak locations on circuits could cause objectionable voltage changes. Possible interaction with other voltage-regulating equipment warrants investigation. This includes other custom-power devices that may be trying to dynamically correct voltage.

Another fast-transfer technology can be used in the same applications as static switches. High-speed mechanical source transfer switches use high-speed vacuum switches and a sophisticated microprocessor based control to provide "break-before-make" transfers in approximately 25 msec or 1.5 cycles. The two sources are not paralleled during the transfer; therefore, the load experiences an interruption of approximately 1.5 cycles. This level of protection may be acceptable for some equipment (but the 1.5-cycle interruption may trip some sensitive equipment). These switches have the advantage of being very efficient (99%), inexpensive (one-fifth to one-tenth the cost of a static switch), and small. Both pole-mounted and padmounted versions are available. For most loads, the fast transfer switch provides significant benefit. Relays and contactors, though, can drop out for a 1.5-cycle interruption.

Normally, fast transfer switches have been applied at individual customers in a primary-selective scheme. The technology could nicely apply to feeder-level application as shown in Figure 2.36. This arrangement provides improved performance for voltage sags, momentary interruptions, and long-duration interruptions for the customers at the end of the circuit (these are customers that usually get the worst power quality).

2.10.3.2 DVRs and Other Custom-Power Devices

In addition to the static transfer switch, medium-voltage power electronics have enabled a wide variety of utility-level solutions to power-quality problems: series injection devices, static regulators, shunt devices, and medium-voltage uninterruptible power supply (UPS) systems with a variety of energy storage options. These power-electronics solutions have been coined "custom power." The advantage of utility-side approaches is that the whole facility is supported (we do not have to track down, test, and fix every possible piece of sensitive equipment).

Figure 2.37 shows the configuration of several custom-power configurations. Most provide support during voltage sags. Single-phase sags are more easily corrected (since devices can use energy from the unfaulted phases). For momentary interruptions, some sort of energy storage is necessary such as batteries, ultracapacitors, flywheel, or superconducting coil. A summary of the cost and capability of the most common devices is shown in Table 2.13.

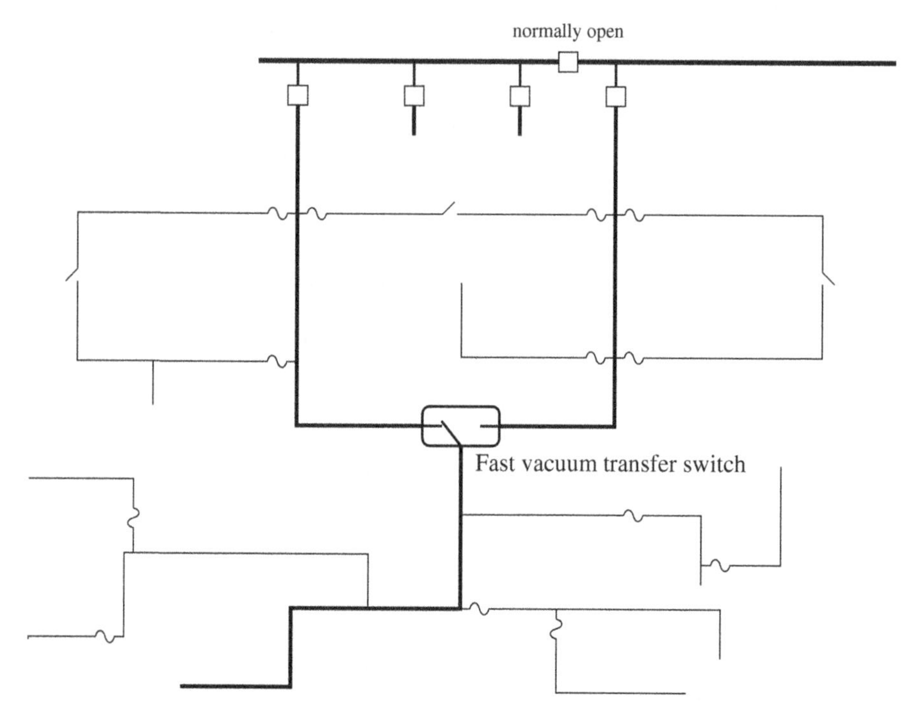

FIGURE 2.36
Using a fast transfer switch to enhance power quality to a downstream section of circuit. (Copyright © 2002. Electric Power Research Institute. 1007281. *Analysis of Extremely Reliable Power Delivery Systems.* Reprinted with permission.)

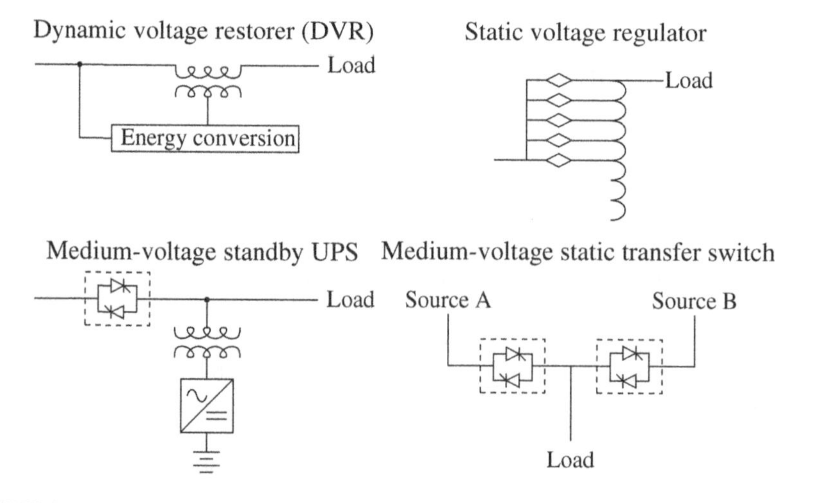

FIGURE 2.37
Common custom-power devices.

TABLE 2.13

Comparison of Custom-Power Equipment for Correcting Sags and Momentary Interruptions

	Cost, U.S. $	Capability
Static shunt compensation	50-200/kvar	sags to 70%
Source transfer switch	500-1000/A	
DVR	150-250/kVA	sags to 50%
Static voltage regulator	80-125/kVA	sags to 50%
MV UPS	750-1500/kVA	

Source: EPRI 1000340, *Guidebook on Custom Power Devices*, EPRI, Palo Alto, CA, 2000.

TABLE 2.14

Comparison of Selected Power Conditioning Equipment

	Sags to 80%	Sags to 50%	Sags to 25%	Below 25%	Outage
Dip-proofing inverter	Solves	Solves	Solves	Solves	To 1 sec
Constant-voltage transformer	Solves	Size[a]	Size[a]	No	No
DySC	Solves	Solves	To 0.33 sec	To 0.26 sec	To 0.15 sec
Uninterruptible power supply	Solves	Solves	Solves	Solves	Solves
Coil hold-in devices	Solves	Solves	Solves	No	No

[a] Size = capability depends on the size of the device

Source: Stephens, M., Johnson, D., Soward, J., and Ammenheuser, J., *Guide for the Design of Semiconductor Equipment to Meet Voltage Sag Immunity Standards*, International SEMATECH, 1999. Technology Transfer #99063760B-TR, available at http://www.sematech.org/public/docubase/document/3760btr.pdf.

The dynamic-voltage restorer (DVR) is one of the devices most suited for correcting for voltage sags (Woodley et al., 1999). During a voltage sag, the DVR adds voltage through an in-line transformer to offset the missing voltage. A DVR can correct for any sag to 50%, and optional energy storage increases the range of performance.

2.10.4 Customer/Equipment Solutions

The best place to provide solutions for customers is often in the facility. Some of the customer solution options are shown in Table 2.14.

Other tips (McEachern, 2001; Stephens et al., 1999) include

- Wire devices in a phase-to-phase configuration where possible.
- Avoid mismatched equipment voltages.
- Avoid the use of ac “ice cube” general-purpose relays.
- Do not use phase-monitoring relays in interlocking circuits.
- Utilize a non-volatile memory.

- Do not overload dc power supplies.
- Use a targeted voltage conditioning approach.
- Use robust inverter drives.

Make sure that computer-controlled equipment can recover from interruptions. This sounds like an obvious solution, but user complaints reveal that a lot of computer-controlled equipment gets completely confused by a disturbance. State machine programming is a promising approach to help make sure that processes recover from disturbances. Another step that helps is to use battery-backed up memory or a disk to store process step information.

2.11 Power Quality Monitoring

Power quality monitoring helps utilities and their customers diagnose and fix power quality problems. Many locations are suitable for monitoring. The distribution substation is a good place because it can monitor the voltage and currents feeding a large number of customers. For a specific customer having a problem, nothing is better than recording right at or near the sensitive process or piece of equipment. Some suggestions for monitoring are as follows:

- *Voltage and current* — Record both voltage and current, and for three-phase installations, record all three phases plus the ground current. For substation monitoring, currents identify which circuit the fault was on or whether it was caused on the transmission system. With the current, we can estimate the distance to the fault. For facility monitoring, the current shows us if a sag is caused by something internal to the facility or on the utility system. More importantly, the current may show what tripped off within the facility and when it tripped.
- *Triggers* — Do not set the triggers too sensitively. Even with larger onboard memories and disk capabilities, there is usually no reason for highly sensitive settings. The commonly used ± 5% is too sensitive. Sensitive triggers create too much data for downloading, storage, and most importantly, for analysis. When recording sags and momentaries, we can safely ignore sags that do not drop below 90%, and a trigger setting of 85% is often appropriate (except for very sensitive loads).
- *Output contacts* — For monitoring at sensitive equipment, record output contacts of any equipment that may indicate its operational status or the process status. Also, be aware that some equipment (like many adjustable-speed drives and UPSs) keeps an event log that may help track down problems. At the substation, record the output contacts of the pertinent relays to determine which relays tripped.

- *Timing* — Synchronize the monitoring clocks to a GPS time reference to help correlate with other monitoring data or event records. If this is not feasible, establish a protocol of regularly setting the recorder clocks (and logging the errors to be able to adjust for drift) to keep them close to a time standard. If a process has internal data recording, synchronize those clocks. The same goes for the utility SCADA system — accurate timing increases the certainty with which we can diagnose power quality problems.
- *Trend data* — Record trend data (say 1-min maximum, minimum, and average values) as well as triggered data (sags, swells, etc.). This helps pinpoint problems related to the steady-state voltage. The ride through of some devices depends on the pre-disturbance voltage.

Fault-recorder type devices with power quality capability are good for substation application; the large number of channels and ability to monitor relay contacts is very beneficial. Also, many types of devices such as relays, recloser controllers, and customer meters can record power quality disturbances.

A drawback to monitoring is that it requires expert manpower to operate. It takes considerable work to install the monitoring, set up the downloading, and (most significantly) analyze the data. Good power quality software helps by highlighting or paring down the data and creating summary reports. Sometimes power quality recorders are used in "stealth" mode: they are installed and largely forgotten until there are customer or utility problems.

An important question is, what do we do with all of the data from power quality recorders? Oftentimes, the best answer is nothing, until power quality problems are reported. A good use of the data is benchmarking. Pick an indicator such as $SARFI_{ITIC}$, and collect data from each recording site. We can compare the benchmark on a site-by-site basis to help determine which circuits need more maintenance attention. Further analysis could reveal what practices are better (for example, circuits with a three-year tree-trimming cycle could be compared with those having a 5-year trim cycle). Benchmarking can also be used as an advertising tool to try and attract sensitive, high-tech customers.

For high accuracy at a given site, we need long monitoring periods. Events that happen less often require longer monitoring for good accuracy. Table 2.15 shows an approximation by Bollen (2000) based on a Poisson distribution. The length of monitoring needed is at least

$$t = \frac{4}{\mu \varepsilon^2}$$

where

ε = accuracy needed, per unit

μ = number of events expected per time period

TABLE 2.15

Monitoring Duration Needed to Achieve the Given Level of Accuracy

Event Frequency	50% Accuracy	10% Accuracy
1 per day	2 weeks	1 year
1 per week	4 months	7 years
1 per month	1 year	30 years
1 per year	16 years	400 years

Source: Bollen, M. H. J., *Understanding Power Quality Problems: Voltage Sags and Interruptions*, IEEE Press, New York, 2000. With permission. ©2000 IEEE.

Treat these as minimum numbers; Bollen's analysis assumes that events randomly appear, while in reality they appear in clusters, during stormy weather or during overloads. This clustering means that even longer monitoring is needed for a given level of accuracy. Another concern is that power systems change over time, which adds further uncertainty to historical monitoring. Bollen also suggests a method of adjusting sag and momentary measurements by the fault rate measured over a longer time period. A better estimate of the actual event rate is

$$\overline{N}_{sags} = N_{sags} \frac{\overline{N}_{faults}}{N_{faults}}$$

where

N_{sags} and N_{faults} = the number of sags and the number of faults over the recording period

$\overline{N}_{sags}$ and $\overline{N}_{faults}$ = the number of sags and the number of faults over a longer period of time

For sites where monitoring cannot meet accuracy needs, prediction methods can help produce an estimate. This requires developing a model of the system and using a stochastic approach to the fault positions analysis method.

References

Abrahams, R., Keus, A. K., Koch, R. G., and van Coller, J. M., "Results of Comprehensive Testing of a 120kW CSI Variable Speed Drive at Half Rating," PQA Conference, 1999.

Bollen, M. H. J., *Understanding Power Quality Problems: Voltage Sags and Interruptions*, IEEE Press, New York, 2000.

Bowes, K. B., "Effects of Power Line Disturbances on Electronic Products," *Power Quality Assurance Magazine*, vol. Premier V, pp. 296–310, 1990.

CEA, *CEA 2000 Annual Service Continuity Report on Distribution System Performance in Electric Utilities*, Canadian Electrical Association, 2001.

Chong, W. Y., "Effects of Power Quality on Personal Computer," Masters thesis, University of Queensland, 2000.

Collins Jr., E. R. and Bridgwood, M. A., "The Impact of Power System Disturbances on AC-Coil Contactors," Textile, Fiber, and Film Industry Technical Conference, 1997.

Conrad, L., Kevin, L., and Cliff, G., "Predicting and Preventing Problems Associated with Remote Fault-Clearing Voltage Dips," *IEEE Transactions on Industry Applications*, vol. 27, pp. 167–72, 1991.

Courtois, E. L., 2001. Personal communication.

Courtois, E. L. and Deslauriers, D., "Voltage Variations Susceptibility of Electronic Residential Equipment," PQA Conference, 1997.

Dorr, D. S., "Point of Utilization Power Quality Study Results," *IEEE Transactions on Industry Applications*, vol. 31, no. 4, pp. 658–66, July/August 1995.

Dorr, D. S., Hughes, M. B., Gruzs, T. M., Juewicz, R. E., and McClaine, J. L., "Interpreting Recent Power Quality Surveys to Define the Electrical Environment," *IEEE Transactions on Industry Applications*, vol. 33, no. 6, pp. 1480–7, November 1997.

EEI, "EEI Reliability Survey," Minutes of the 8th Meeting of the Distribution Committee, March. 28–31, 1999.

EPRI 1000340, *Guidebook on Custom Power Devices*, EPRI, Palo Alto, CA, 2000.

EPRI 1001665, *Power Quality Improvement Methodology for Wires Companies*, Electric Power Research Institute, Palo Alto, CA, 2003.

EPRI 1007281, *Analysis of Extremely Reliable Power Delivery Systems: A Proposal for Development and Application of Security, Quality, Reliability, and Availability (SQRA) Modeling for Optimizing Power System Configurations for the Digital Economy*, Electric Power Research Institute, Palo Alto, CA, 2002.

EPRI PEAC Brief No. 7, *Undervoltage Ride-Through Performance of Off-the-Shelf Personal Computers*, EPRI PEAC, Knoxville, TN, 1992.

EPRI PEAC Brief No. 9, *Low-Voltage Ride-Through Performance of 5-hp Adjustable-Speed Drives*, EPRI PEAC, Knoxville, TN, 1993.

EPRI PEAC Brief No. 10, *Low-Voltage Ride-Through Performance of AC Contactor Motor Starters*, EPRI PEAC, Knoxville, TN, 1993.

EPRI PEAC Brief No. 12, *Low-Voltage Ride-Through Performance of a Modified Personal Computer Power Supply*, EPRI PEAC, Knoxville, TN, 1993.

EPRI PEAC Brief No. 17, *Electronic Digital Clock Performance During Steady-State and Dynamic Power Disturbances*, EPRI PEAC, Knoxville, TN, 1994.

EPRI PEAC Brief No. 30, *Ride-Through Performance of Adjustable- Speed Drives with Flying Restart*, EPRI PEAC, Knoxville, TN, 1995.

EPRI PEAC Brief No. 39, *Ride-Through Performance of Programmable Logic Controllers*, EPRI PEAC, Knoxville, TN, 1996.

EPRI PEAC Brief No. 44, *The Effects of Point-on-Wave on Low-Voltage Tolerance of Industrial Process Devices*, EPRI PEAC, Knoxville, TN, 1998.

EPRI PEAC commentary #3, *Performance of AC Motor Drives During Voltage Sags and Momentary Interruptions*, 1998.

EPRI TP-113781, *Reliability Benchmarking Application Guide for Utility/Customer PQ Indices*, Electric Power Research Institute, Palo Alto, CA, 1999.

EPRI TR-106294-V2, *An Assessment of Distribution System Power Quality: Volume 2: Statistical Summary Report*, Electric Power Research Institute, Palo Alto, CA, 1996.

EPRI TR-106294-V3, *An Assessment of Distribution System Power Quality: Volume 3: Library of Distribution System Power Quality Monitoring Case Studies*, Electric Power Research Institute, Palo Alto, CA, 1996.

IEEE Std. 493-1997, *IEEE Recommended Practice for the Design of Reliable Industrial and Commercial Power Systems (Gold Book).*

IEEE Std. 1159-1995, *IEEE Recommended Practice for Monitoring Electric Power Quality.*

IEEE Std. 1366-2000, *IEEE Guide for Electric Power Distribution Reliability Indices.*

Information Technology Industry Council (ITI), "ITI (CBEMA) Curve Application Note," 2000. Available at http://www.itic.org.

Kojovic, L. A. and Hassler, S. P., "Application of Current Limiting Fuses in Distribution Systems for Improved Power Quality and Protection," *IEEE Transactions on Power Delivery*, vol. 12, no. 3, pp. 791–800, April 1997.

Kojovic, L. A., Hassler, S. P., Leix, K. L., Williams, C. W., and Baker, E. E., "Comparative Analysis of Expulsion and Current-Limiting Fuse Operation in Distribution Systems for Improved Power Quality and Protection," *IEEE Transactions on Power Delivery*, vol. 13, no. 3, pp. 863–9, July 1998.

Mansoor, A., Collins, E. R., Bollen, M. H. J., and Lahaie, S., "Effects of Unsymmetrical Voltage Sags on Adjustable Speed Drives," PQA-97, Stockholm, Sweden, June 1997.

McEachern, A., "How to Increase Voltage Sag Immunity," 2001. Available at http://powerstandards.com/tutorials/immunity.htm.

Sabin, D. D., Grebe, T. E., and Sundaram, A., "RMS Voltage Variation Statistical Analysis for a Survey of Distribution System Power Quality Performance," IEEE/PES Winter Meeting Power, February 1999.

Sekine, Y., Yamamoto, T., Mori, S., Saito, N., and Kurokawa, H., "Present State of Momentary Voltage Dip Interferences and the Countermeasures in Japan," International Conference on Large Electric Networks (CIGRE), September 1992.

SEMI F42-0600, *Test Method for Semiconductor Processing Equipment Voltage Sag Immunity*, Semiconductor Equipment and Materials International, 1999.

SEMI F47-0200, *Specification for Semiconductor Processing Equipment Voltage Sag Immunity*, Semiconductor Equipment and Materials International, 1999.

Short, T. A., Mansoor, A., Sunderman, W., and Sundaram, A., "Site Variation and Prediction of Power Quality," IEEE Transactions on Power Delivery, 2002. Preprint submission.

St. Pierre, C., "Don't Let Power Sags Stop Your Motors," *Plant Engineering*, pp. 76–80, September 1999.

Stephens, M., Johnson, D., Soward, J., and Ammenheuser, J., *Guide for the Design of Semiconductor Equipment to Meet Voltage Sag Immunity Standards*, International SEMATECH, 1999. Technology Transfer #99063760B-TR, available at http://www.sematech.org/public/docubase/document/3760btr.pdf.

Williams, B. R., Schmus, W. R., and Dawson, D. C., "Transmission Voltage Recovery Delayed by Stalled Air Conditioner Compressors," *IEEE Transactions on Power Systems*, vol. 7, no. 3, pp. 1173–81, August 1992.

Woodley, N. H., Morgan, L., and Sundaram, A., "Experience with an Inverter-Based Dynamic Voltage Restorer," *IEEE Transactions on Power Delivery*, vol. 14, no. 3, pp. 1181–6, July 1999.

All I know is that when shit hits the fan, management calls us we don't call them.

In response to: Do linemen feel they are respected by management and coworkers for the jobs they are doing, do management and coworkers understand what you do?

www.powerlineman.com

3

Other Power Quality Issues

While voltage sags and momentary interruptions cause the most widespread power quality problems, several other power quality disturbances can damage equipment, overheat equipment, disrupt processes, cause data loss, and annoy and upset customers. In this chapter, we explore several of these, including transients, harmonics, voltage flicker, and unbalance.

3.1 Overvoltages and Customer Equipment Failures

Often, customers complain of equipment failures, especially following power interruptions. Is it lightning? Voltage swells during faults? Some sort of switching transient? Sometimes explanations are obvious, sometimes not. Several events can fail equipment during a fault/interruption, either from the disturbance that caused a fault, the voltage sag during the fault, a voltage swell during the fault, or the inrush while the system is recovering. Some possibilities are

- *Overvoltages* — Lightning and other system primary-side overvoltages can enter the facility and damage equipment.
- *Grounding* — Poor facility grounding practices can introduce overvoltages at equipment from fault current.
- *Capacitive coupling* — Reclose operations and other switching transients can create fast-rising voltage on the primary that capacitively couples through the transformer, causing a short pulse on the secondary.
- *Inrush current* — While recovering from a voltage sag or momentary interruption, the inrush current into some electronic equipment can blow fuses or fail semiconductor devices.
- *Unbalanced sags* — Three-phase electronic equipment like adjustable-speed drives can draw excessive current during a single-phase sag

or other unbalanced sag. The current can blow fuses or fail the front-end power electronics.

- *Equipment aging* — Some equipment is prone to failure during turn on, even without a voltage transient. The most obvious example is an incandescent light bulb. Over time, the filament weakens, and the bulb eventually fails, usually when turned on. At turn on, the rapid temperature rise and mechanical stress from the inrush can break the filament.

Lightning can cause severe overvoltages, both on the primary and on the secondary. Damaging surges can enter from strikes to the primary, strikes to the secondary, strikes to the facility, strikes to plumbing, and strikes to cable-television or telephone wires. Poor grounding practices can make lightning-caused failures more likely.

Another source of severe overvoltages is primary or secondary conductors contacting higher voltage lines. Other overvoltages are possible; normally these are not severe enough to damage most equipment, except for sensitive electronics:

- Voltage swells — Peaks at about 1.3 per unit on most distribution circuits.
- Switching surges — Normally peaks at less than 2 per unit and decays quickly.
- Ferroresonance — Normally peaks at less than 2 per unit.

Just as arresters on distribution lines are sensitive to overvoltages, arresters inside of electronic equipment often are the first thing to fail. The power supply on most computers and other electronics contains small surge arresters (surge suppressors) that can fail quickly while trying to clamp down on overvoltages, especially longer-duration overvoltages. These small suppressors have limited energy absorption capability.

In addition to proper grounding, surge arresters are the primary defense against lightning and other transients. For best protection, use surge protection at the service entrance and surge protection at each sensitive load.

Surge arresters work well against short-duration overvoltages — lightning and switching transients. But arresters have trouble conducting temporary power-frequency overvoltages; they absorb considerable energy trying to clamp the overvoltage and can fail. Small arresters often are the first component to fail in equipment. Using a higher voltage rating helps give more protection to the surge arrester during temporary overvoltages (for example, end users should not use arresters with a maximum continuous operating voltage below 150 V). Surge arresters should be coordinated; the large surge arrester at the service entrance should have the lowest protective level of all of the arresters within the facility. Because arresters are so nonlinear, the unit with the lowest protective level will conduct almost all of the current.

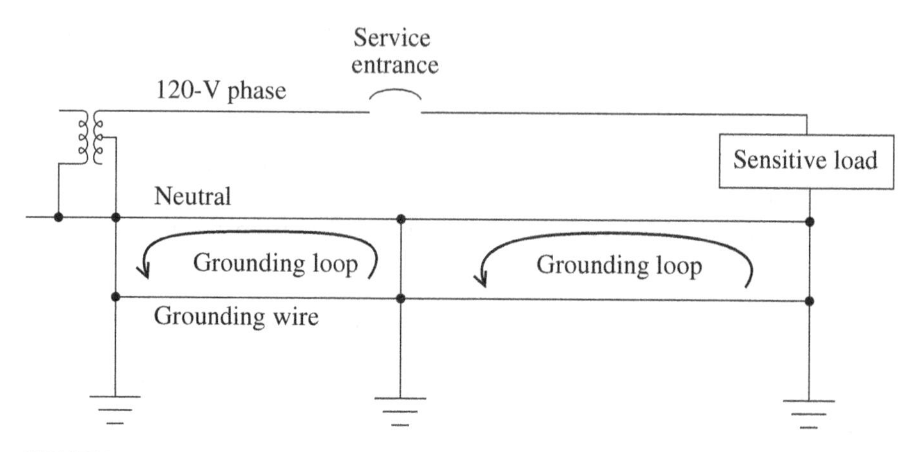

FIGURE 3.1
Ground loops within a facility.

So, we want the arrester with the most energy capability to absorb most of the energy.

3.1.1 Secondary/Facility Grounding

Grounding problems within a facility can lead to equipment failures or malfunctions. A common problem is ground loops. If the secondary neutral has multiple connections within the facility, external ground currents from lightning or from faults can induce voltages along the neutral. Figure 3.1 shows an example. The neutral voltage shifts can impose overvoltages on sensitive equipment. For more information, see IEEE Std. 1100-1999 or IEEE Std. 142-1991. Single-point grounding of the neutral breaks the loops and reduces noise and possible overvoltages.

Grounding loops between multiple "ports" can create damaging scenarios. Any significant wired connection to equipment counts as a port — power, telephone, cable TV, printer cables, and networking cables. Figure 3.2 shows an example of a television where the cable enters via a different route and has a different ground than its electric service. A fault or lightning strike on the utility side can elevate the potential of the electric-supply ground relative to the television cable ground. Even if both the electric and cable television cables have independent surge protection, surges still create voltage differences between components. To avoid multiple-port difficulties, arrange to have all power and communications enter a building at one location, provide a common ground, and apply surge protection to all ports. To overcome this problem, users can install a "surge reference equalizer" rather than independent surge protection on each port (EPRI PEAC Solution No. 1, 1993). Both the electric supply and the other port (telephone, cable TV, ethernet, etc.) plug into the surge reference equalizer. The equalizer provides surge protection for both incoming ports and a single common grounding connection.

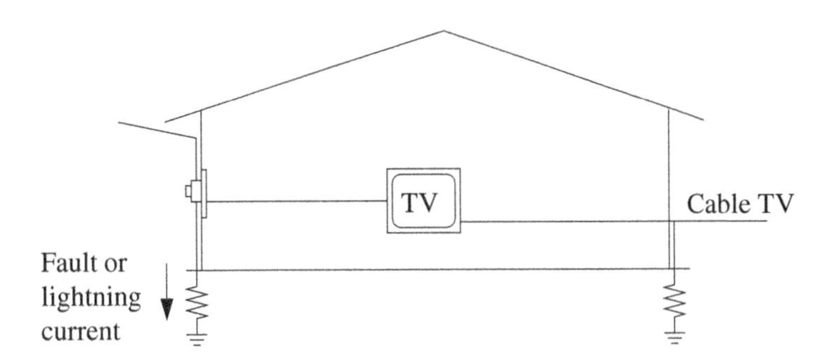

FIGURE 3.2
Example of a two-port voltage stress at a television between its power source and the cable TV cable.

For equipment, overvoltages between the phases and neutral are most important. While it is commonly believed that neutral-to-ground overvoltages cause problems for sensitive electronics, EPRI PEAC found that computers are quite tolerant of neutral-to-ground overvoltages. In tests, neither a continuous 50-V neutral-to-ground voltage nor a 3-kV, 100-kHz ring wave upset the operation of a computer (EPRI PEAC Brief No. 21, 1994).

Pole-mounted controllers for regulators, reclosers, capacitors, and switches face particularly severe environments. Because of their location, they have significant exposure to transients from lightning strikes to nearby poles. Additionally, they are right in the path of fault current returning from faults downstream of the controller. Some utilities have had reliability problems with controllers; some problems may stem from poor powering and grounding arrangements. Figure 3.3 shows how significant voltages can develop on the low-voltage supply when the power is supplied at a remote pole and a fault occurs downstream. Lightning current following this same path can create very severe voltages. Additionally, two-port vulnerabilities can arise

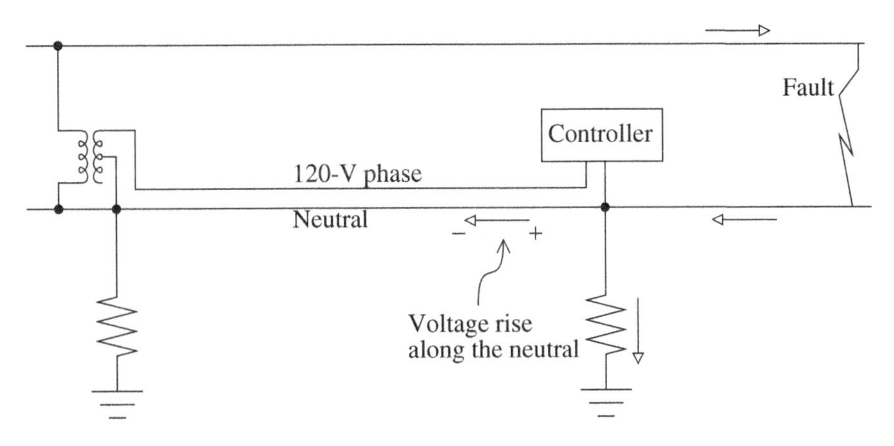

FIGURE 3.3
Pole-mounted controller exposure to faults along the circuit.

between the controller power and potential transformers (PTs). Other sources of two-port problems are current transformers and communication lines. Consider the following strategies to better protect these controllers:

- *Surge protection* — On the power-supply inputs, apply extra low-voltage surge protection, especially in high lightning areas.
- *Power source* — To help avoid two-port problems, try to power the controller from a PT or small transformer on the same structure as the controller.

3.1.2 Reclose Transients

One source of possibly damaging surges is from capacitive coupling through transformers. The very short-duration pulses come from electrostatic coupling through the transformer. Much analysis of this has been done for larger power transformers at generating stations (Abetti et al., 1952). At high frequencies, the transformer acts like a capacitor. A steep surge can pass from the primary to the secondary. How much of the surge gets from the primary to the secondary depends on the capacitances of the transformer and the secondary load — not the transformer turns ratio. A transformer has capacitance from the high-voltage winding to the low-voltage winding (C_1) and capacitance from the secondary winding to ground (C_2). As a first approximation, the voltage on the secondary as a function of the voltage on the primary is a capacitive voltage divider (Greenwood, 1991):

$$V_S = \frac{C_1}{C_1 + C_2} V_P$$

Line energization and capacitor switching and other transients can create surges that pass through the transformer. It does not even have to be an overvoltage on the primary, just a fast rise to the nominal peak — a normal reclose operation — can create a surge on the secondary. When a line is energized near its peak voltage, right when the switch engages, a traveling wave with a very steep front rushes down the line. At a transformer, this front can be steep enough to couple capacitively through to the secondary at voltages much higher than the turns ratio of the transformer. Figure 3.4 shows a 1-μsec wide transient to just over 2000 V following a reclose operation.

These capacitively coupled surges are worse with

- Higher voltage transformations from the primary to the secondary (34.5 kV to 480 V is worse than 12.5 kV to 480 V).
- Loads close to the substation or recloser (the wave front flattens out with distance).
- Minimal resistive load on the secondary.

Millisecond-scale view of the transient:

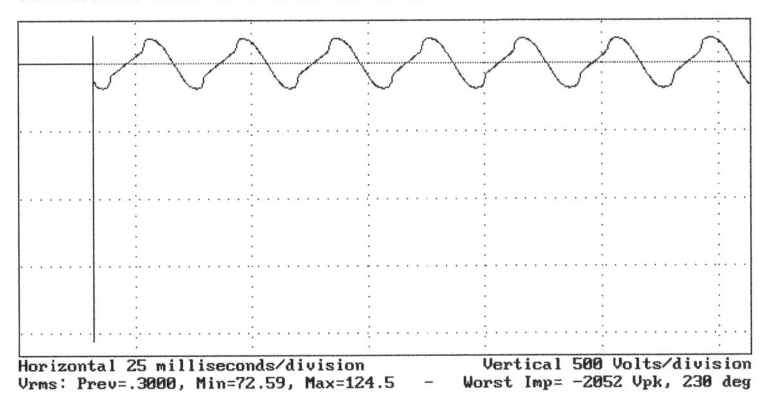

Microsecond-scale view:

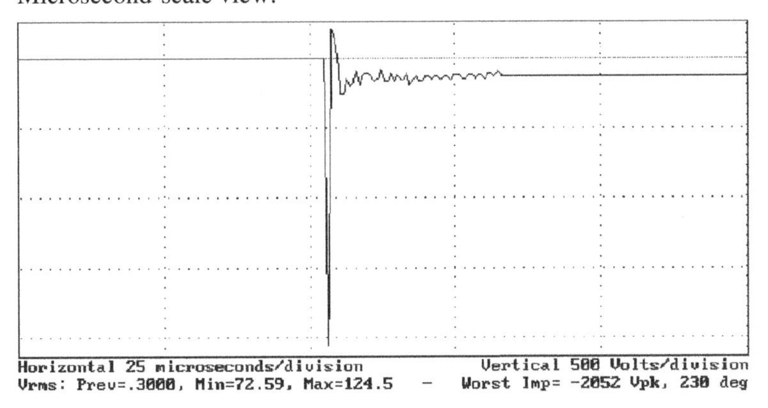

FIGURE 3.4
Measurements of a transient captured during a circuit reclose. (Recordings courtesy of Francisco Ferrandis Mauriz, Iberdrola Distribucion Electrica.)

We can protect against these surges with:

- Line-to-ground capacitor banks in the facility (either power-factor correction or surge capacitors)
- Surge arresters

3.2 Switching Surges

Transients are triggered from capacitor switching, from line energization, and from faults. Capacitor switching transients normally cause the highest peak magnitudes. If a capacitor is switched just when the system voltage is near its peak, the capacitor pulls the system voltage down (as current rushes into the capacitor to charge it up). The system rebounds; the voltage overshoots and oscillates about the fundamental-frequency waveform (theoretically rising to two per unit, but normally less than that). This transient normally decays quickly. The oscillation occurs at the natural resonance frequency between the capacitor bank and the system, usually in the hundreds of hertz:

$$f = \frac{1}{2\pi\sqrt{L_S C}} = 60\sqrt{\frac{X_C}{X_S}} = 60\sqrt{\frac{MVA_{sc}}{Mvar}}$$

where

f = frequency, Hz
C = capacitance, F
L_S = system inductance up to the capacitor bank at nominal frequency, H
X_C = line-to-ground impedance of one phase of the capacitor bank at nominal frequency
X_L = system impedance up to the capacitor bank at nominal frequency
MVA_{SC} = three-phase short-circuit MVA at the point where the capacitor is applied
$Mvar$ = three-phase Mvar rating of the capacitor

Capacitor switching transients tend to be more severe when the capacitor is at a weaker point on the system. Other capacitors on the circuit normally help, but if two capacitors are very close together, switching in the second capacitor can create a higher voltage transient. Capacitors with significant separation can magnify the switching surges.

Normally, switching surges are not particularly severe on distribution systems. The voltages decay quickly, and magnitudes are normally not severe enough to fail line equipment (Figure 3.5 shows a typical example). Switching transients can be large enough to affect sensitive end-use loads, particularly adjustable-speed drives and uninterruptible-power supplies. The oscillation frequency is in the hundreds of hertz, low enough for the transient to pass right through distribution transformers and into customers' facilities.

EPRI's distribution power quality (DPQ) study found regular but mild transient overvoltages, most presumably due to switching operations. Transients measured on the distribution primary above 1.6 per unit are rare, averaging less than two per year (where per unit is relative to the peak of the nominal sinusoidal voltage wave). Figure 3.6 shows the average distribution of transients measured during the study (EPRI TR-106294-V2, 1996). This graph shows occurrences of transients with a peak magnitude between 1.05 and 1.9 per unit with a principal oscillation frequency between 240 and 3000 Hz (excluding transients associated with faults).

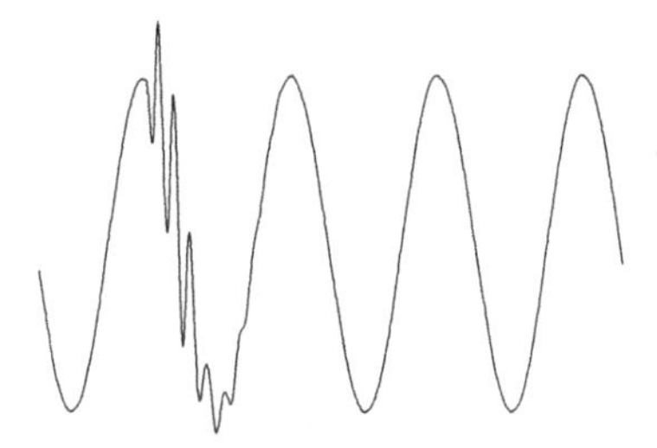

FIGURE 3.5
Capacitor switching transient. (Copyright © 1996. Electric Power Research Institute. TR-106294-V2. *An Assessment of Distribution System Power Quality: Volume 2: Statistical Summary Report.* Reprinted with permission.)

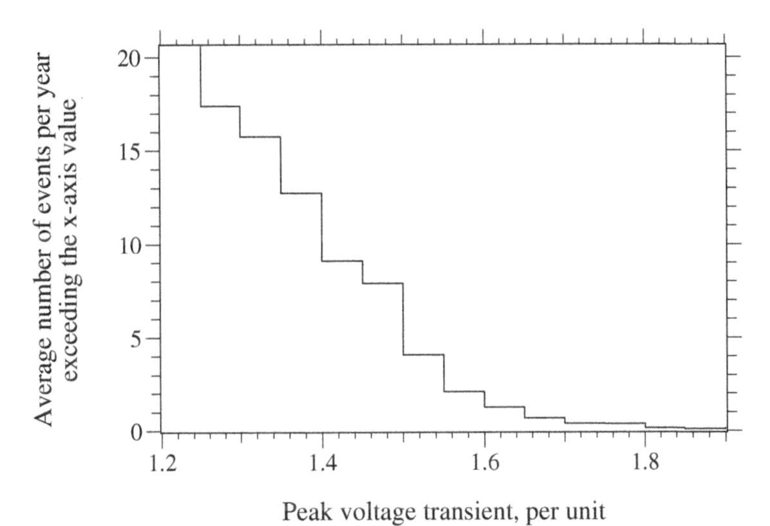

FIGURE 3.6
Average magnitudes of voltage transients measured on the distribution primary in EPRI's DPQ study. (Data from [EPRI TR-106294-V2, 1996; Sabin et al., 1999b].)

During EPRI's DPQ study, most of the transients measured occurred during the morning hours, from 5 a.m. to 10 a.m., when most switched capacitors are coming on line. The wide majority of oscillating transients measured in the study had dominant frequencies between 250 and 800 Hz.

AEP staged various switching tests on a 12.47-kV circuit and found that most capacitor switching and line energization transients were less than two per unit on the primary (Kolcio et al., 1992). In their tests, switching a 450-kvar capacitor bank produced the highest transient, just under 2.3 per unit on closing — vacuum switches produced slightly higher transients than oil switches. Line energizations caused peak primary voltage of just under 1.9 per unit. The transients decayed in less than 5 msec.

Switching transients are normally more problematic on higher voltage distribution circuits, such as 34.5 kV. Several instances of equipment failures have been reported on these higher voltage systems, usually from switching a large (10 plus Mvar) capacitor bank (Shankland et al., 1990; Troedsson et al., 1983). Capacitors tend to be larger, and sources are weaker. Plus, at higher voltages, primary level insulation capability is lower on a relative basis compared to lower voltage systems.

In capacitive circuits, switch *restrike* or *prestrike* generates more severe transients. A restrike can occur when switching a capacitor off; the switch opens at a current zero, trapping the peak voltage on the capacitor (see Figure 3.7). As the system voltage wave decreases from the peak, the voltage across the switch rapidly increases. By the time the system voltage reaches the opposite system peak, the voltage across the switch is 2 per unit (this extra voltage is the main reason that switches have a different rating for breaking capacitive circuits). If the switch restrikes because of this overvoltage, the voltage swings about the new voltage with a peak-to-peak voltage of 4 per unit, forcing the line-to-ground voltage to a peak of 3 per unit (theoretically). The interrupter can clear at another current zero and trap even more voltage on the capacitor, which can cause another more severe restrike; such multiple restrikes can escalate the voltage. Fortunately, such voltage escalation is rare. Restrikes are most likely if the interrupting contacts have not fully separated when the current first interrupts. Restrikes can fail switches and cause higher voltages on the system. Most switching technologies, including vacuum, can restrike under some conditions.

Prestriking occurs when a switch closes into a capacitor. If the gap between a switch's contacts breaks down before the contacts have fully closed, the system voltage charges up the capacitor. Because the contacts are not closed, the switch can clear at a current zero, leaving the capacitor charged. We are

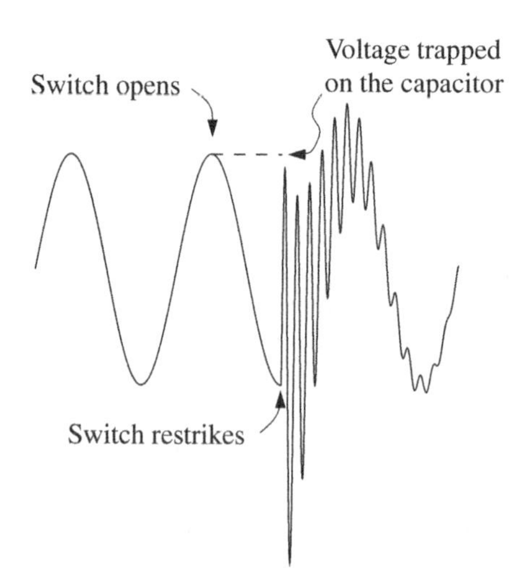

FIGURE 3.7
Restrike of a capacitor bank.

back to the restriking scenario — when the gap breaks down again, the voltage transient can oscillate around the system voltage at a much higher level. Capacitor switches should not and normally do not regularly restrike or prestrike, and voltages rarely escalate, but some failures can be traced to these switching problems.

Capacitor switching is not the only switching transient that can occur on a distribution circuit. Line energization can cause transients, as can faults. Both of these are normally benign. Walling et al. (1995) tested load-break elbows and switching-compartment disconnects. Such switching results in repeated arc restrikes and prestrikes, but the overvoltages are not particularly severe because the oscillations dampen out without clearing at a transient zero crossing (no escalation of the voltage). With a maximum overvoltage of 2.44 per unit, Walling et al. do not expect consequential effects on underground cable insulation.

3.2.1 Voltage Magnification

Certain circuit resonances can *magnify* capacitor switching transients at locations away from the switched capacitor bank (Schultz et al., 1959). Magnification often increases a transient's peak magnitude by 50%, sometimes even by 100%. Figure 3.8 shows the classic voltage magnification circuit; two sets of resonant circuits interact to amplify the transient that occurs when C_1 is switched. The most severe magnification occurs when:

- C_1 is much larger than C_2.
- The series combination of L_1 and C_1 resonates at about the same frequency as L_2 and C_2.
- There is little series or shunt resistance to dampen the transient.

If C_1 is much larger than the series impedance $L_2 + C_2$, switching C_1 in will initiate a transient that oscillates at a frequency of $\omega_1 = 1/\sqrt{L_1C_1}$. If the downstream pair, L_2 and C_2, happen to have a natural resonant frequency

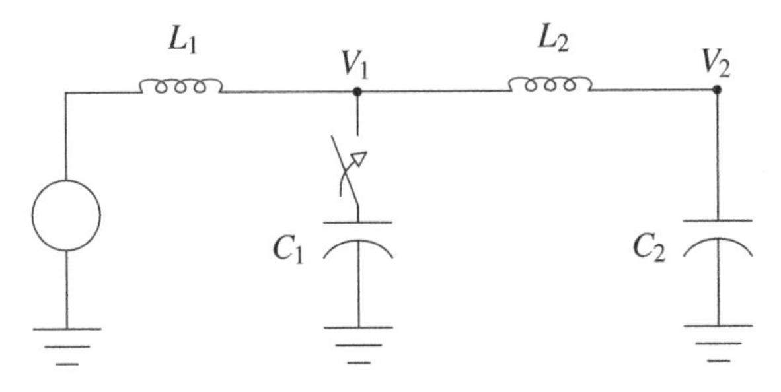

FIGURE 3.8
Voltage magnification circuit.

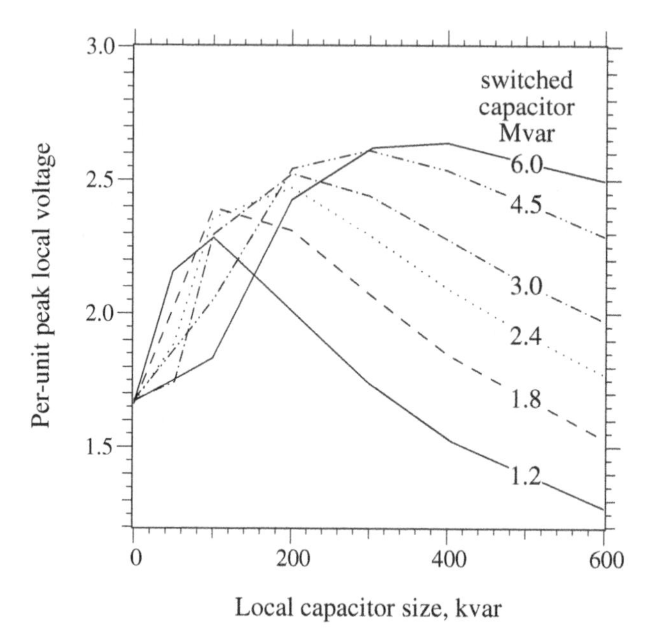

FIGURE 3.9
Transient magnitude for various sizes of switched distribution feeder capacitors and low-voltage capacitors. (From McGranaghan, M. F., Zavadil, R. M., Hensley, G., Singh, T., and Samotyj, M., "Impact of Utility Switched Capacitors on Customer Systems — Magnification at Low Voltage Capacitors," *IEEE Trans. Power Delivery*, 7(2), 862-868, April 1992. With permission. ©1992 IEEE.)

near that of ω_1, oscillation between L_1 and C_1 will pump the natural oscillation between L_2 and C_2, causing higher voltages at V_2.

Consider a 10-Mvar substation capacitor and a 1200-kvar line capacitor on a 12.5-kV circuit with a substation source impedance of 1 Ω at 60 Hz. Under these conditions, the loop resonances match when X_{L2} is 8.3 Ω, or when the capacitor is about 13 mi from the substation (given typical line impedances). For maximum magnification, the ratio of the inductive impedances equals the inverse of the capacitor kvar ratios ($kvar_1/kvar_2 = X_{L2}/X_{L1}$).

In another common scenario, low-voltage power factor correction capacitors magnify transients from switched utility capacitors (McGranaghan et al., 1992). In this scenario, L_2 is primarily determined by the customer's transformer, and L_1 is determined by the system impedance to the capacitor bank. Their parametric study found that a wide combination of local and switched utility capacitors can magnify transients as shown in Figure 3.9. Larger utility banks are more likely to generate transients that are magnified on the low-voltage side. Resistive load helps dampen these transients; facilities with primarily motor loads have higher transients. Magnified transients can also inject considerably more energy into low-voltage metal-oxide arresters than they are typically rated to handle (especially for switching large utility banks).

Transients from switched capacitor banks on the high-side of the substation bus can magnify at distribution capacitors or customer-owned capacitors (Bellei et al., 1996). Again, a wide combination of capacitor ranges can magnify surges.

3.2.2 Tripping of Adjustable-Speed Drives

The main power quality symptoms related to distribution capacitor switching is shutdown of adjustable-speed drives (ASD) and other sensitive process equipment. The front-end of an adjustable-speed drive rectifies the incoming line-to-line ac voltages to a dc voltage. The rectifier peak-tracks the three incoming ac phases, so a switching surge on any phase charges up the dc bus. Because the electronics fed by the dc bus are sensitive to overvoltages, drives normally have sensitive dc bus overvoltage trip settings, typically 1.2 per unit with some as low as 1.17 per unit. It does not take much of a transient to reach 1.2 per unit. At such low sensitivities, the system and customer voltage regulation plays an important role. If the voltage at the drive is at 1.05 per unit when the capacitor switches, the transient does not have to oscillate much to reach 1.17 or 1.2 per unit. These voltages are on the dc bus, which is fed by the line-to-line voltage. The actual line-to-ground transients on the distribution system normally need to be higher than 1.2 per unit to raise the drive dc link voltage to 1.2 per unit (and this depends on the customer transformer connection).

McGranaghan et al. (1991) used transient simulations to analyze several variables impacting the tripping of drives:

- *dc capacitor* — If the drive's rectifier has a dc capacitor, it impacts the drive's sensitivity to switching transients. Drives with larger dc capacitor take longer to charge up, which makes them less likely to trip on overvoltage.
- *Switch closing* — The worst transients are when the capacitor switches all switch at the peak of the waveform.
- *Capacitor size* — Larger capacitor banks cause a more severe transient. For the case McGranaghan et al. (1991) analyzed, capacitor banks less than 1200-kvar banks were less likely to cause drive tripouts (see Figure 3.10).
- *Local power-factor correction capacitors* — If the facility has local capacitors, magnification of the incoming transient makes the drive more likely to trip. End users can avoid this problem by configuring their power-factor correction capacitors as tuned filters (McGranaghan et al., 1992).

Locally, the addition of chokes (series inductors) connected to the input terminals of the adjustable-speed drive reduces the voltage rise on the drive's

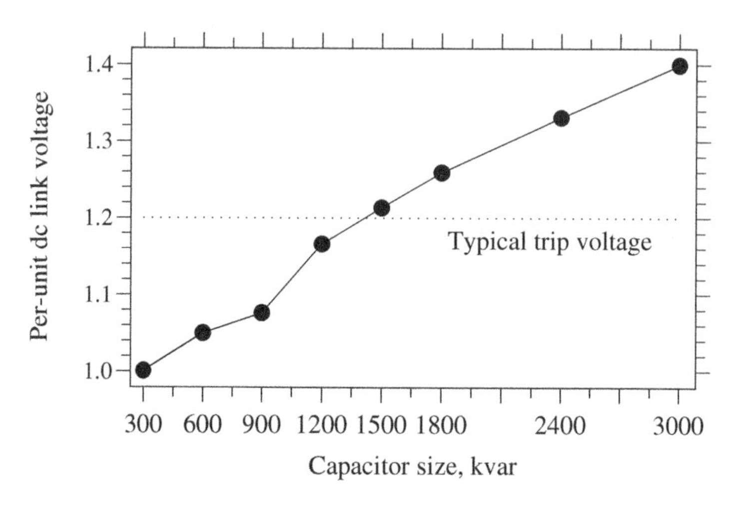

FIGURE 3.10
dc voltage on an adjustable-speed drive from switching of the given size capacitor bank (with no local power-factor correction capacitors). (From McGranaghan, M. F., Grebe, T. E., Hensley, G., Singh, T., and Samotyj, M., "Impact of Utility Switched Capacitors on Customer Systems. II. Adjustable-Speed Drive Concerns," *IEEE Trans. Power Delivery*, 6(4), 1623-1628, October 1991. With permission. ©1991 IEEE.)

dc bus. Standard sizes are 1.5, 3, and 5% impedance on the drive's kVA rating. A parametric simulation study of transmission-switched capacitors found that a 3% inductor greatly reduced the number of drive trips, but a number of transients can still trip some drives (depending on the size capacitor switched and many other variables) (Bellei et al., 1996).

3.2.3 Prevention of Capacitor Transients

Switched capacitors cause the most severe switching transients. Utility-side solutions to prevent the transient include zero-voltage closing switches or pre-insertion resistors or reactors. Zero-voltage closing switches time their closing to minimize transients. Switches with pre-insertion resistors place a resistor across the switch gap before fully closing contacts; this reduces the inrush and dampens the transient. Any of these options are available for substation banks, and utilities use them often, especially in areas serving sensitive commercial or industrial customers.

For distribution feeder banks, the zero-crossing switch controllers are the only available option for reducing transients. These controllers time the closing of each switch contact to engage the capacitor at the instant where the system voltage is zero, or very close to zero. An uncharged capacitor closing in at zero volts causes no transient.

The switches time the initiation of the switching so that by the time the switch engages, the system voltage is at zero. Since the capacitor should have no voltage, and it is switched into the system when it has no voltage, the switching does not create a transient. Timing and repeatability are important. One popular 200-A vacuum switch takes 0.015 sec to close and has a repeatability of ±3 degrees (±0.14 msec). Each phase is controlled separately.

Other claimed benefits of zero-crossing switches include increased capacitor and switch life, reduction of induced voltages into the low voltage control wiring, and reduction of ground transients. Zero-crossing switches also eliminate capacitor inrush, including the much more severe inrush found when switching a capacitor with a nearby capacitor already energized.

3.3 Harmonics

Distortions in voltage and current waveshapes can upset end-use equipment and cause other problems. Harmonics are a particularly common type of distortion that repeats every cycle. Harmonically distorted waves contain components at integer multiples of the base or *fundamental* frequency (60 Hz in North America). Resistive loads like incandescent lights, capacitors, and motors do not create harmonics — these are passive elements — when applied to 60-Hz voltage; they draw 60-Hz current. Electronic loads, which create much of the harmonics, do not draw sinusoidal currents in response to sinusoidal voltage.

A very common harmonic producer is the power supply for a computer, a switched-mode power supply that rectifies the incoming ac voltage to dc (see Figure 2.21). The bridge rectifier has diodes that conduct to charge up capacitors on the dc bus. The diodes only conduct when the ac supply voltage is above the dc voltage for just a portion of each half cycle. So, the power supply draws current in short pulses, one each half cycle. Each pulse charges up the capacitor on the power supply. The current is heavily distorted compared to a sine wave, containing the odd harmonics, 3, 5, 7, 9, etc. The third harmonic may be 80% of the fundamental, and the fifth may be 60% of the fundamental. Figure 3.11 shows examples of the harmonic current drawn by switched-mode power supplies in computers along with other sources of harmonics.

Other very common sources of harmonics are adjustable-speed drives and other three-phase dc power supplies (see Figure 2.28). These rectify the incoming ac waveshape from each of the three phases. In doing so, drives create current with harmonics of order 5, 7, 11, 13, etc. — all of the odd harmonics except multiples of three. In theory, drives create a fifth harmonic that is one-fifth of the fundamental, a seventh harmonic that is one-seventh of the fundamental, and so forth. Other harmonic-producing loads include arc furnaces, arc welders, fluorescent lights (with magnetic and especially with electronic ballasts), battery chargers, and cycloconverters.

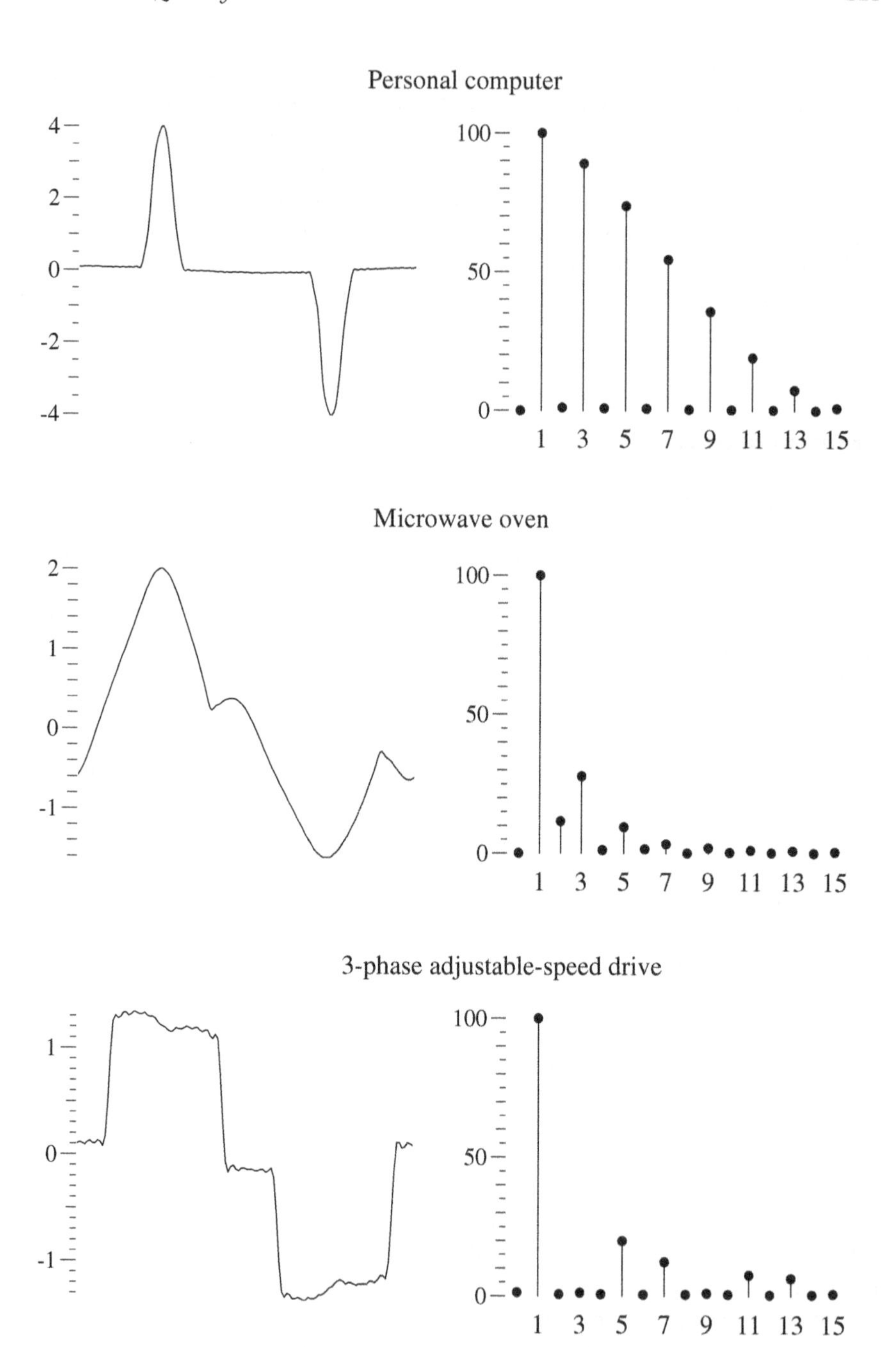

FIGURE 3.11
Examples of harmonic-current sources. (Data from EPRI PEAC, various measurements.)

Harmonic currents and harmonic voltages are interrelated. Commonly, nonlinear loads are modeled as ideal current injections ("ideal" meaning they inject the same current regardless of the voltage). An *n*th-harmonic source driving current through an impedance $R+jX$ raises the *n*th voltage to

$$V_n = R \cdot I_n + jn \cdot X \cdot I_n$$

Utility distribution systems are normally stiff enough (low Z) that they can absorb significant current distortions without voltage distortion.

Harmonics have been heavily researched and studied, probably more so than any other power quality problem. Despite the attention, harmonic problems are uncommon on distribution circuits. Most harmonic problems are isolated to industrial facilities; end-use equipment produces harmonics, which cause other problems within the facility.

We can characterize harmonics by the magnitude and phase angle of each individual harmonic. Normally, we specify each harmonic in percent or per unit of the fundamental-frequency magnitude. Another common measure is the *total harmonic distortion*, THD, which is the square-root of the sum of the squares of each individual harmonic. Voltage THD is

$$V_{THD} = \frac{\sqrt{\sum_{h=2}^{\infty} V_h^2}}{V_1}$$

where V_1 is the rms magnitude of the fundamental component, and V_h is the rms magnitude of component h. The Fourier transform is used to convert one or more waveshape cycles to the frequency domain and V_h values.

IEEE provides a recommended practice for harmonics, IEEE Std. 519-1992, which gives limits for utilities and for end users. Utilities are expected to maintain reasonably distortion-free voltage to customers. For suppliers at 69 kV and below, IEEE voltage limits are 3% on each individual harmonic and 5% on the total harmonic distortion (Table 3.1). Both of these percentages are referenced to the nominal voltage.

Current distortion limits (Table 3.2) were developed with the objective of matching the voltage distortion limits. If users limit their current injections according to the guidelines, the voltage limits will remain under the guidelines imposed on the utility (as long as no major circuit resonance exists). The individual harmonics and the total harmonics (total demand distortion, TDD) are referenced to the maximum demand at that point, either the 15- or 30-min demand. Harmonic current limits depend on the facility size relative to the utility at the interconnection point. Larger facilities — relative to the strength of the utility at the interconnection point — are more restricted. Likewise, a facility at a stronger point in the system is allowed to

TABLE 3.1

Voltage Distortion Limits

	Voltage Distortion Limit
Individual harmonics	3%
Total harmonic distortion (THD)	5%

Note: For a bus limit at the point of common coupling at and below 69 kV. For conditions lasting less than 1 h, the limits may be exceeded by 50%.

Source: IEEE Std. 519-1992.

TABLE 3.2

Current Distortion Limits for Distribution Systems (120 V through 69 kV)

I_{sc}/I_L	$h < 11$	$11 \leq h < 17$	$17 \leq h < 23$	$23 \leq h < 35$	$34 \leq h$	TDD
< 20	4.0	2.0	1.5	0.6	0.3	5.0
20–50	7.0	3.5	2.5	1.0	0.5	8.0
50–100	10.0	4.5	4.0	1.5	0.7	12.0
100–1000	12.0	5.5	5.0	2.0	1.0	15.0
> 1000	15.0	7.0	6.0	2.5	1.4	20.0

Note: Even harmonics are limited to 25% of the odd harmonic limits above. Current distortions that result in a dc offset are not allowed. For conditions lasting less than 1 h, the limits may be exceeded by 50%.

Source: IEEE Std. 519-1992.

inject more than the same facility at a weaker point because the same injection at a weaker point creates more voltage distortion. The ratio I_{sc}/I_L determines which set of limits applies. I_{sc} is the maximum short-circuit current; I_L is the maximum demand for the previous year.

Both the voltage and current limits apply at the *point of common coupling*, the point on the system where another customer can be served. Either side of the distribution transformer can be the point of common coupling, depending on whether the transformer can supply other customers. The point of common coupling is often taken as the metering point.

Harmonics can impact a variety of end-use equipment (IEEE Task Force, 1985; IEEE Task Force, 1993; Rice, 1986). Some of the most common end-use effects are as follows:

- *Transformers* — Current distortion increases transformer losses and heating, enough so that transformers may have to be derated [see (IEEE Std. C57.110-1998)]. Dry-type transformers are particularly sensitive.
- *Motors* — Voltage distortion induces heating on the stator and on the rotor in motors and other rotating machinery. The rotor presents a relatively low impedance to harmonics, like a shorted transformer

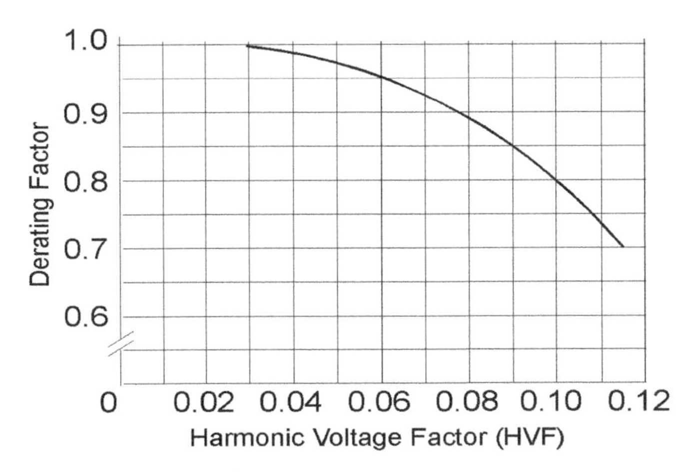

FIGURE 3.12
Motor derating due to harmonics. NEMA's harmonic voltage factor is slightly different from THD; it includes only the odd harmonics that are not multiples of three. (From NEMA MG-1, *Standard for Motors and Generators*, 1998. With permission.)

winding. The effects are similar to motors operating with voltage unbalance, although not quite as severe since the impedance to the harmonics increases with frequency. Figure 3.12 shows the National Electrical Manufacturers Association's (NEMAs) recommended derating factor for motors running with distorted voltage.

- *Conductors and Cables* — Resistance increases with frequency due to skin effect and proximity effect, particularly on larger conductors. As an example, at the seventh harmonic (420 Hz), a 500-kcmil cable has a resistance that is 2.36 times its dc resistance (Rice, 1986). Neutral conductors in facilities are especially prone to problems; third-harmonic currents from single-phase loads add in the neutral, which can actually increase neutral current above that in the phase conductors (EPRI PEAC Application #6, 1996).
- *Sensitive electronics* — Ironically, some of the main harmonic producers are also sensitive to harmonics. If an uninterruptible power supply (UPS) detects excessive distortion, it may switch to its batteries; after the batteries run out, the critical load may drop out. Harmonics can impact digital devices that depend on a voltage zero crossing for timing (digital clocks being the simplest example). Excessive harmonics can introduce additional zero crossings that can disrupt controllers.

Most harmonic problems occur in industrial or commercial facilities. Distribution equipment is fairly immune to problems from harmonic voltages or currents. Heating from harmonics is not normally a problem for overhead conductors. Underground cables are more sensitive to heating. Harmonic currents can appreciably increase cable temperatures.

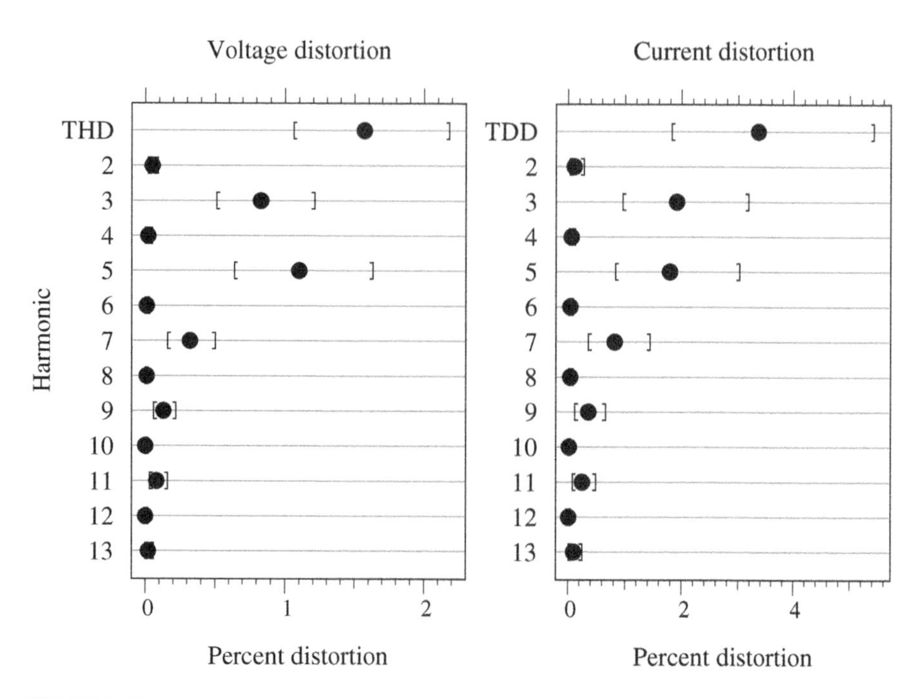

FIGURE 3.13
Voltage and current distortion measured in EPRI's DPQ study. The dot marks the mean distortion for the given harmonic, and the brackets mark the 5th and 95th percentiles. (Data from [EPRI TR-106294-V2, 1996; Sabin et al., 1999a].)

Many harmonic problems relate to heating, either in transformers, motors, neutral conductors, or capacitors. Because of the thermal time constants of equipment, heat-related problems occur for sustained levels of harmonics (usually along with heavy power-frequency loading). Heating degrades insulation and leads to premature failure. Higher peak voltages can also cause insulation breakdowns, with capacitors being highly sensitive.

Most distribution circuits have very little distortion, having neither voltage distortion nor current distortion. Figure 3.13 shows average distortion measured during EPRI's DPQ study (EPRI TR-106294-V2, 1996; Sabin et al., 1999a). None of the DPQ sites had an average voltage THD above the IEEE 519 limit of 5% (the worst couple of sites averaged just under 4.8% THD). And, only 4% of the DPQ sites exceeded this IEEE 519 limit more than 5% of the time. At the DPQ sites, 95% of the time, the voltage THD was less than 2.2%.

Substation sites had slightly lower voltage and current distortions than those recorded at feeder sites: substation average THD = 1.37%; overall average THD = 1.57%. Industrial feeders (defined as having at least 40% of the load coming from industrial loads) showed only slightly higher voltage distortion than the average for all sites.

Other monitoring studies also show fairly low harmonics. Surveys of 76 sites on the Southwestern Electric Power Company's system found that none exceeded a voltage THD of 5% (less than 5% of exceeded 4%; 87% had THD less than 2.5%) (Govindarajan et al., 1991). A few of the sites had current distortion exceeding IEEE 519 standards. In a large survey of locations on the Sierra Pacific Power Company, voltage THD averaged 1.59%, and current THD averaged 5.3% (Etezadi-Amoli and Florence, 1990).

Some work has been done to investigate the impact of new harmonic loads on the harmonic distortion. Compact fluorescent lights with electronic ballasts, heat pumps and air conditioners with adjustable-speed drives, computers, and electric vehicle battery chargers are some of the loads that could drive up harmonics. Pileggi et al. (1993) predicted that if every home used two or three electronically ballasted fluorescent lights, voltage distortion may exceed 5%. This is hard to believe, but it does make us consider the impact of future electronic loads. Dwyer et al. (1995) also came to similar conclusions: modest numbers of electronic compact fluorescent lights (50 W per house) can raise the voltage distortion above 5%, especially if feeder capacitors create resonances that amplify the harmonics. It helps that multiple single-phase loads have some phase-angle cancellation; harmonic loads do not necessarily sum linearly (Mansoor et al., 1995).

In the northeastern U.S., Emanuel et al. (1995) predict rising levels of harmonic distortion based on predictions of increased use of adjustable-speed drives, electronic fluorescent lights, computer loads, and other nonlinear loads. They predict that more nonlinear loads can add as much as 0.3% per year to the voltage THD under the worst-case scenario. A limited number of measurements by the same group of researchers over a 10-year time period found that harmonic voltages have increased at a rate of about 0.1% per year (it takes 10 years for the THD to increase by a percentage point) (Nejdawi et al., 1999).

Penetrations of air conditioners driven by adjustable-speed drives to 10 to 20% can raise voltage THD levels above 5% (Gorgette et al., 2000; Thallam et al., 1992). In another study, Bohn (1996) predicted that electric-vehicle battery charger penetration levels as low as 5% could increase voltage THD above 5%, especially on weak systems. In both of these cases, investigators found wide differences in equipment designs. Some generate much more harmonics than others. For example, some adjustable-speed drives with more sophisticated rectifier sections cause much less harmonics than standard rectifiers. As electronic loads grow, utilities must involve themselves in setting guidelines to limit the harmonics introduced by end-use equipment.

Finding the source of harmonics can be tricky — IEEE 519A provides two ways to track down harmonics (IEEE P519A draft 7, 2000):

1. *Time variations* — If harmonics are intermittent, correlating the time variations of the voltage or current harmonics with the operation time of facilities or of specific equipment can identify the harmonic producer.
2. *Monitoring with capacitor banks off* — In a radial circuit with no capacitors, harmonic sources inject current that flows back to the power

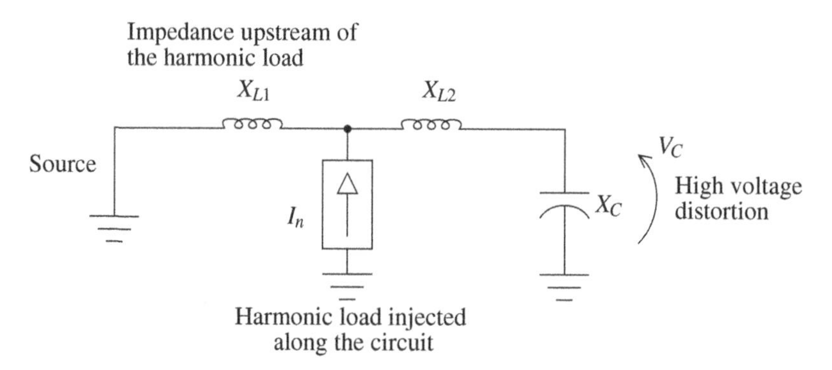

FIGURE 3.14
Harmonic resonance.

source; following the current leads to the source of the harmonics. If capacitors are present, resonances can mask the true source of harmonics.

3.3.1 Resonances

When harmonics cause problems, capacitors are often a contributing factor. Capacitors can cause resonances that amplify harmonics. Harmonic problems often show up first at capacitor banks, where harmonics cause fuse operations or even capacitor failure. Most problems with severe harmonics are found in industrial facilities where capacitors resonate against the system impedance. Utility resonances are infrequent but sometimes cause problems.

A capacitor on a distribution circuit will resonate against the inductance back to the system source (including the line impedances and transformer impedances). At the resonant frequency, the circuit amplifies harmonics injected into the circuit (see Figure 3.14).

The resonance point between the capacitor and the system for the scenario in Figure 3.14 (this is the same frequency that the system will ring at during a switching surge) is

$$n = \sqrt{\frac{X_C}{X_L}} = \sqrt{\frac{MVA_{SC}}{Mvar}}$$

where

n = order of the harmonic (multiples of the fundamental frequency)
X_C = line-to-ground reactance of one phase of the capacitor bank at nominal frequency
$X_L = X_{L1} + X_{L2}$ = system reactance at nominal frequency
MVA_{SC} = three-phase short-circuit MVA at the point where the capacitor is applied
$Mvar$ = three-phase Mvar rating of the capacitor

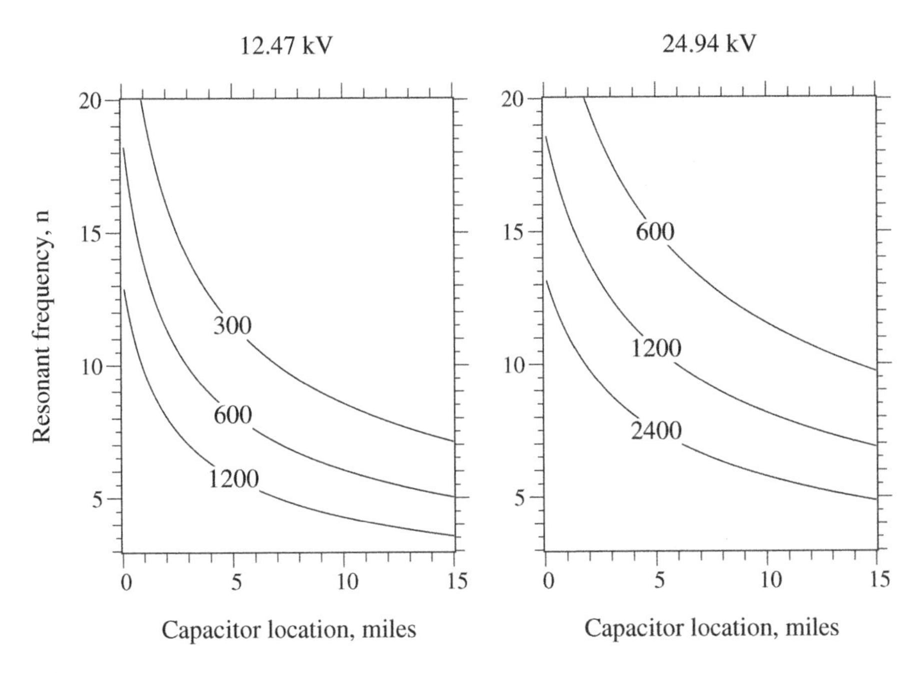

FIGURE 3.15
Resonant frequency vs. capacitor size (in kvar) and location for balanced, positive-sequence resonances (500-kcmil AAC).

If a nonlinear load injects a harmonic frequency equal to the system's resonant point, it can create significant harmonic voltages. Common problem frequencies for resonances are $n = 5$, 7, 11, and 13. Larger capacitors lower the resonant point to where it is more likely to cause problems. For 15-kV class systems, resonances are possible at common capacitor locations for 600 to 1200-kvar banks (see Figure 3.15). For the single-capacitor circuit in Figure 3.14, the voltage distortion at the capacitor is:

$$V_C = \frac{X_C X_{L1}}{nX_{L1} + nX_{L2} - \frac{1}{n}X_C} I_n$$

where I_n is the nth harmonic current. The worst conditions are when the harmonic source is right at the capacitor or downstream of the capacitor (X_{L2} is small). Harmonics injected further upstream are amplified less. This simplification ignores the resistance of the line and the resistance of the loads. Both reduce the peak magnitude.

Multiple capacitors on a circuit create multiple resonant points that can require more sophisticated analysis. For complicated circuits, especially with capacitor banks, a harmonic analysis program helps identify problems and try out solution options. A harmonic program models multiple harmonic sources at different locations; the solution shows the harmonic current flows

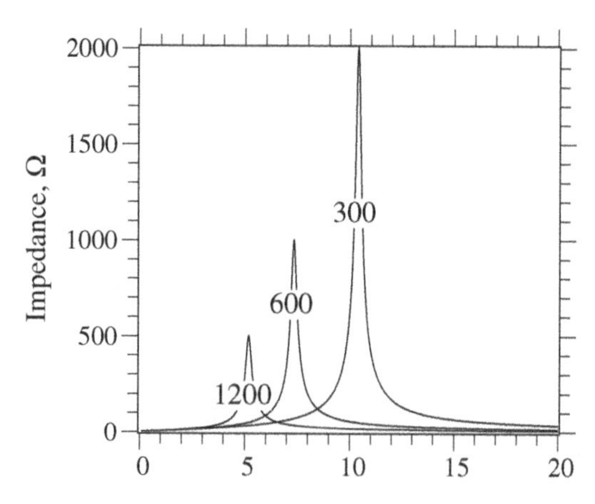

FIGURE 3.16
Harmonic scans for various size capacitors (shown in kvar) applied on a 12.47-kV circuit 6 miles from the substation.

and the harmonic voltages. Another tool, the frequency scan — a plot of the impedance at a location vs. frequency — helps identify system resonances that appear as peaks in the scan (see Figure 3.16).

Distribution systems rarely need solutions to harmonics because typical voltage distortion levels are so low. When problems arise, specific harmonic producers or specific resonant conditions are often to blame. For offending producers, we can enforce harmonic standards. Where resonances occur, the solutions include:

- *Relocate or remove capacitors* — The easiest solution is to disconnect the capacitors that form the resonant circuit. Relocating capacitors or changing their size can also shift the resonant point enough to ease problems.
- *Harmonic filters* — While rarely done, tuned filters, consisting of a capacitor in series with an inductor, absorb harmonics near their tuned frequency. The best location for filters is closest to the harmonic producers. If harmonic producers are distributed, filters located near the center of the line can absorb some of the harmonics. Applying filters requires study to ensure that additional resonances are not created.

3.3.2 Telephone Interference

Current flowing through distribution conductors induces voltage along parallel conductors, both power conductors and communication conductors. On low-voltage communication lines, induced voltages can interfere with

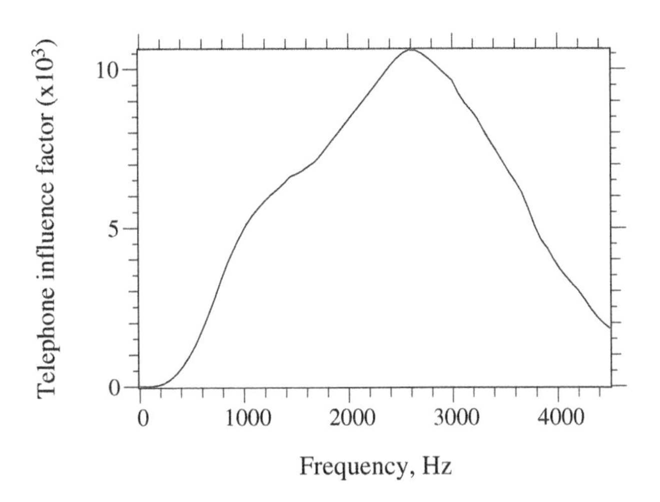

FIGURE 3.17
Telephone interference factor weights (W_f) with frequency.

communication signals, particularly on telephone circuits. Telephone interference is less common than in the past, primarily because telephone companies use more immune circuits with shielded cables or fiber optics. Open-wire communication lines are the most susceptible to interference. Interference depends on the offending current magnitude, the distance between conductors, and the length that the circuits run in parallel.

Harmonics increase the likelihood of telephone interference. Higher frequencies introduce noise that transfers more readily through telephone equipment and is more easily heard. *Telephone influence factors (TIF)* weight harmonics differently, depending on frequency. Figure 3.17 shows TIF weighting factors with frequency. This curve is based on the response of the telephone equipment (mainly the handset) and human perception of different frequencies. The most sensitive frequency is near 2600 Hz (n = 43). Lower-order harmonics scaling as far down as the third harmonic can interfere with telephone and other communication circuits. The total TIF for either voltage or current is a weighted factor at each harmonic relative to the total rms current:

$$TIF = \sqrt{\sum \left(\frac{X_f W_f}{X_t} \right)^2}$$

where

X_t = total rms voltage or current
X_f = single-frequency rms current or voltage at frequency f
W_f = single-frequency TIF weighting at frequency f

TABLE 3.3

Balanced $I \cdot T$ Guidelines between Overhead Power Circuits and Overhead Telephone Circuits

Classification	$I \cdot T$
Levels most unlikely to cause interference	≤ 10,000[a]
Levels that might cause interference	10,000–25,000
Levels that probably will cause interference	> 25,000

[a] For some areas that use a ground return for either telephone or power circuits, this value may be as low as 1500.

Source: IEEE Std. 519-1992.

TIF is a factor that indicates the harmonic content weighted by frequency; TIF is like THD but scales each harmonic according to its relative interference capability. Because interference depends on current magnitudes as well as harmonic content, interference potential is often described in terms of the product of current and the telephone influence factor, $I \cdot T$, where I is the rms current in amperes, and T is the TIF. Table 3.3 shows guidelines for $I \cdot T$ from IEEE 519.

Harmonics in the ground return path are much more likely to interfere with communication circuits. Balanced third and ninth harmonics as well as other multiples of three (the *triplens* or *triples*) are zero-sequence frequencies. These currents generally add in the neutral. Because zero-sequence circuit impedances are higher than positive-sequence impedances, zero-sequence resonances can arise with smaller capacitor banks than for positive-sequence resonances. Figure 3.18 shows capacitor locations that cause zero-sequence resonances that force more current into the ground-return path. Table 3.4 shows locations where capacitors resonate for zero-sequence third harmonics.

While the triplen harmonics are most likely to interfere with communications circuits, other common harmonics such as the fifth or seventh can cause trouble. Although the fifth and seventh harmonics are not naturally zero-sequence currents, if the harmonic components on each of the three phases are unbalanced, the unbalanced portion of harmonics will flow in the ground.

Telephone interference problems are usually solved by the telephone company in cooperation with the electric utility involved. Solutions are often trial-and-error; harmonic load-flow models can help test out solutions. The normal utility-side solutions are (IEEE Std. 776-1992; IEEE Std. 1137-1991; IEEE Working Group, 1985):

- *Change size* — For problems involving a resonance, increasing or decreasing the size of a bank can shift the resonant point enough to ease interference problems. The easiest solution is to disconnect the bank that is contributing to the problem. This is also a good first step to quantify the role of the capacitor bank in the interference.

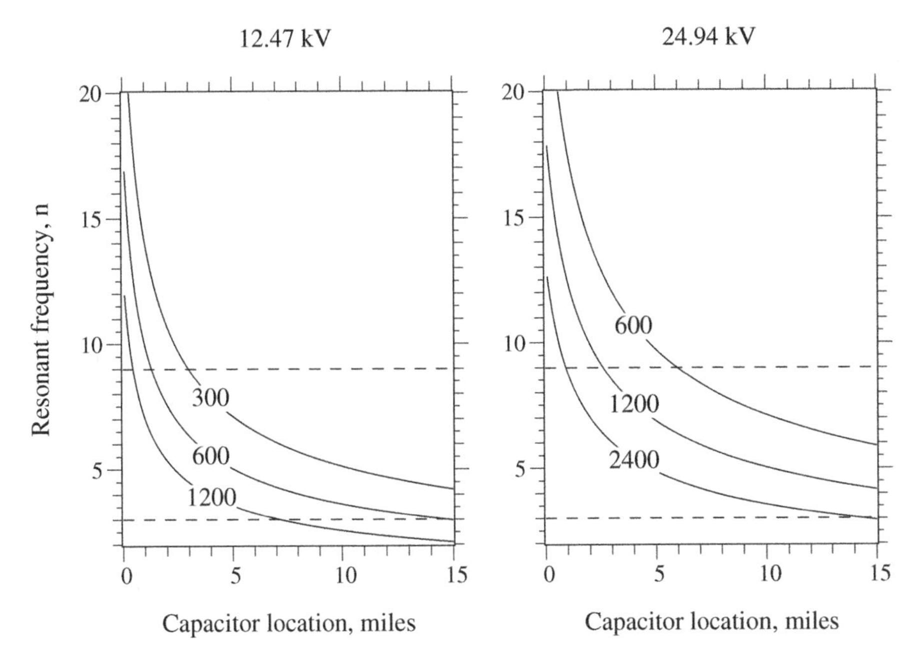

FIGURE 3.18
Resonant frequency vs. capacitor size (in kvar) and location for zero-sequence resonances (500-kcmil AAC, $X_0 = 1.9\Omega/\text{mile}$).

TABLE 3.4
Capacitor Location in Miles from the Substation Where a 3rd-Harmonic Zero-Sequence Resonance Occurs for the Given Size Capacitor (for 500-kcmil AAC with $X_0 = 0.3605\Omega/1000$ ft)

	System Voltage, kV			
kvar	**4.8**	**12.47**	**24.94**	**34.5**
600	2.1	14.8	59.8	
900	1.3	9.7	39.6	76.2
1200	1.0	7.2	29.5	56.9
1800		4.7	19.4	37.6
2400			14.4	27.9
4800			6.8	13.4

- *Balancing* — Balancing harmonic loads between phases reduces the ground return current. In cases of long single-phase runs, we can upgrade the line to a two-phase line to reduce ground currents. Note that this will not help with the third-harmonic currents.
- *Move the bank* — Moving a capacitor can change a resonant point enough to stop interference problems. On some circuits, one can also move the capacitor away from the telephone circuits having problems.

- *Unground the bank* — A floating-wye connection has no connection to the ground, so the connection blocks zero-sequence harmonic currents. Two-bushing capacitor units are necessary for floating the wye point (unless the utility floats the capacitor tanks and deals with the safety issues that accompany that practice).
- *Add a neutral filter* — While not a common solution, a tuned reactor in the ground path of a wye-connected capacitor bank is invisible to positive and negative sequences, but it changes the zero-sequence resonant frequency of a distribution feeder and often eliminates the resonant problem. As another approach, on a bank close to the harmonic source, one could also tune the filter to drain the offending harmonics (so they do not flow back along the electric/phone lines).
- *Add a grounding bank* — Grounding banks can change zero-sequence impedances to shift resonances or shift current flows away from parallel utility/communication runs to avoid interference. They also provide a low-impedance path for zero-sequence characteristic currents to flow. A zig-zag bank can be added, or existing floating-wye – delta connections can be converted to grounded wye – delta.

On the communications side, several solutions are possible: conversion to more immune circuits such as fiber optics or shielded conductors. 60-Hz rejection filters, drainage coils between conductor pairs to reduce induced voltages, and neutralizing transformers which add a voltage opposing the induced voltages. Finally, if the interference can be traced to major harmonic producers, the offending producer can be turned off or harmonics can be filtered at the source.

3.4 Flicker

Light flicker is due to rapidly changing loads that cause fluctuation in the customer's voltage. Even a small change in voltage can cause noticeable lamp flicker. Flicker is an irritation issue. Flicker does not fail equipment; flicker does not disrupt sensitive equipment; flicker does not disrupt processes. Flickering lights or televisions or computer monitors annoy end users.

Annoying lamp flicker occurs when rapid changes in load current cause the power system voltage to fluctuate. Sawmills, irrigation pumps, arc welders, spot welders, elevators, laser printers — all can rapidly change their current draw. In arc furnaces, arcs fluctuate wildly from cycle to cycle, continuously flickering the voltage. When starting, motors draw significant inrush, five or six times their normal current, possibly depressing the voltage for tens of seconds. Some loads like elevators turn on and off repeatedly. All of these fluctuating load changes can cause flicker.

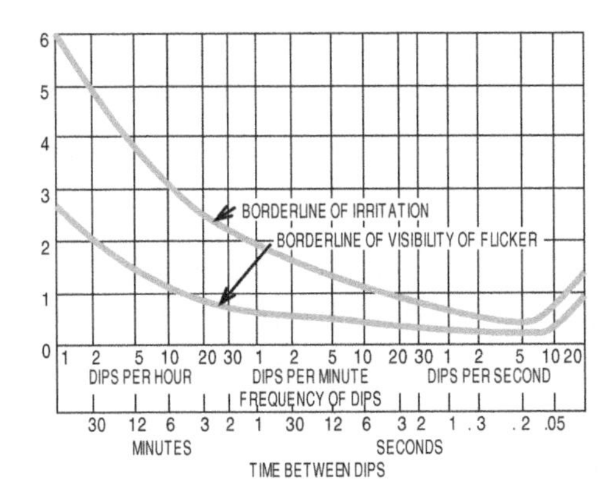

FIGURE 3.19
GE flicker curve. (From IEEE Std. 141-1993. Copyright 1994 IEEE. All rights reserved.)

Susceptibility to flicker depends on the stiffness of the supply system. So, flicker is more common on lower-voltage systems and at the ends of long circuits.

Both magnitude and frequency of fluctuations affect flicker perceptions. People are most sensitive to flicker that changes from two to ten times per second, and flicker is visible up to about 35 Hz. The most common flicker reference curve that has been developed is the GE flicker curve based on tests by General Electric and several utilities around 1930 (General Electric GET-1008L) and republished in the IEEE Red Book (IEEE Std. 141-1993). [See Walker (1979) for comparison with other flicker curves.] An IEEE survey found that 69% of utilities were using the GE curve (Seebald et al., 1985). The GE curve shows both a "threshold of perception" and a "threshold of irritation" (see Figure 3.19). The GE flicker curve is based on square-wave changes to the supply voltage at the frequencies indicated. Load changes that are more gradual than a stepped square wave result in less noticeable flicker; the eye–brain response is more sensitive to rapid light changes (up to a point). The flicker curve is based on a change in voltage (ΔV) relative to the steady-state voltage (V) (see Figure 3.20). Both ΔV and V are best represented by rms quantities; mixing rms, peak, or peak-to-peak quantities gives errors of 2 or $\sqrt{2}$. The frequency of the changes is also confusing, even for changes that are regular and periodic. The GE flicker curve is based on the number of *dips* per minute or second or hours. Some other curves or criteria are based on the number of *changes* per unit of time (frequency of changes are twice the frequency of dips).

The flicker tendency of different lights varies significantly:

- *Smaller incandescent bulbs* — Smaller filaments cool more quickly, so their light output changes more for a given fluctuation. Higher-voltage bulbs also have smaller filaments, so they flicker less.

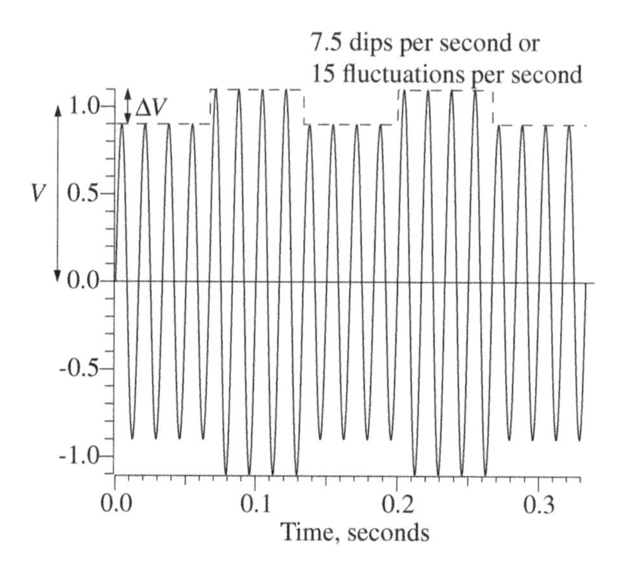

FIGURE 3.20
Characterizing the magnitude and frequency of voltage changes.

- *Dimmers* — Some electronic dimmers greatly increase voltage flicker, especially at low-light settings (EPRI PEAC Brief No. 25, 1995). Flicker on some dimmed lights is equivalent to the same flicker on undimmed lights with double the voltage fluctuation.
- *Fluorescent lights* — Whether magnetic or electronically ballasted, fluorescent lights normally flicker less for a given voltage input, some by a factor of over five (voltage changes on a fluorescent light need to be five times greater to cause the same level of flicker perception on an incandescent light). This is not always the case though; some fluorescent lights flicker as much as incandescent lights, and there is no general rule to identify whether a fluorescent light is sensitive. Fluorescent light ouput depends on the point on wave where the phosphors ignites, so peak voltage affects fluorescent lights more than rms-voltage changes.

While lights receive the most attention, voltage fluctuations can also cause televisions and computer monitors to waver.

Because flicker is based on human perception, it is an inexact science. Some people are more sensitive than others. Younger people are more sensitive. Flicker is more noticeable in quiet, low-light settings. In addition, the steady-state voltage during the time of the fluctuation impacts flicker — lower nominal voltages lead to less flicker; a 5% reduction in voltage reduces perceived flicker by 5% (EPRI PEAC Brief No. 36, 1996). Flicker is more noticeable with side vision (peripheral vision) than with straight-ahead vision. The perception process involves the thermal characteristics of the filament and the human eye–brain response.

The European community has derived sophisticated methods of characterizing and analyzing flicker, developing a flickermeter (IEC 868, 1986) and a comprehensive set of standards. The IEEE is moving towards adopting the European flickermeter approach to quantify flicker (Halpin et al.). The flickermeter models the complex lamp–eye–brain interaction to uniformly quantify flicker from a variety of sources, whether the offending load is an arc furnace or repetitive motor starts. The flickermeter measures voltage and produces these outputs:

- *Instantaneous flicker sensation* — This output is in "perception units," a unit-less quantity. One per unit is defined as the threshold of perception for 50% of the population.
- *Short-term flicker indicator* (P_{st}) — The short-term flicker indicator is a weighted factor based on probabilities of the instantaneous flicker sensation over a ten-minute monitoring period. A $P_{st} = 1$ is the threshold of irritation, the level that the majority of the population finds annoying. P_{st} values as low as 0.7 have been found to be visible under some conditions.
- *Long-term flicker indicator* (P_{lt}) — Long-term flicker is based on 2 h of flicker measurement and combines 12 consecutive P_{st} values (using the cubed root of the sum of the cubes of the 12 P_{st} values).

The flickermeter produces comparable results to the GE flicker curve when the flickermeter has a square-wave input as shown in Figure 3.21. Both the GE curve and the IEC curve (calibrated for 120-V lamps) produce comparable results. A flicker curve analysis is easier for loads that fluctuate periodically in a square-wave fashion. For intermittent or chaotic loads like welders, a flicker curve is impossible to use. Converting a measured waveform to the frequency domain with a Fourier transform and plotting the results on a flicker curve is imprecise. While the exercise reveals predominant frequencies, the Fourier transform does not mimic the brain's response to flickering light. The flickermeter is much more appropriate for cases where loads fluctuate irregularly.

The P_{st} is an rms quantity that we can treat like a per-unit voltage. If the current draw from the fluctuating load doubles (and the waveshape stays the same), the voltage deviation (ΔV) will double, and P_{st} will double. Just as we can use voltage dividers, we can use impedance dividers to estimate flicker propagation and estimate flicker severity at other locations on a circuit (see Figure 3.22). Flicker decreases as one moves upstream of the fluctuating load on a radial circuit. Customers at or downstream of the fluctuating load experience the worst flicker.

Multiple sources of harmonics can influence flicker differently, depending on the characteristics of the fluctuating loads. Multiple sources combine as

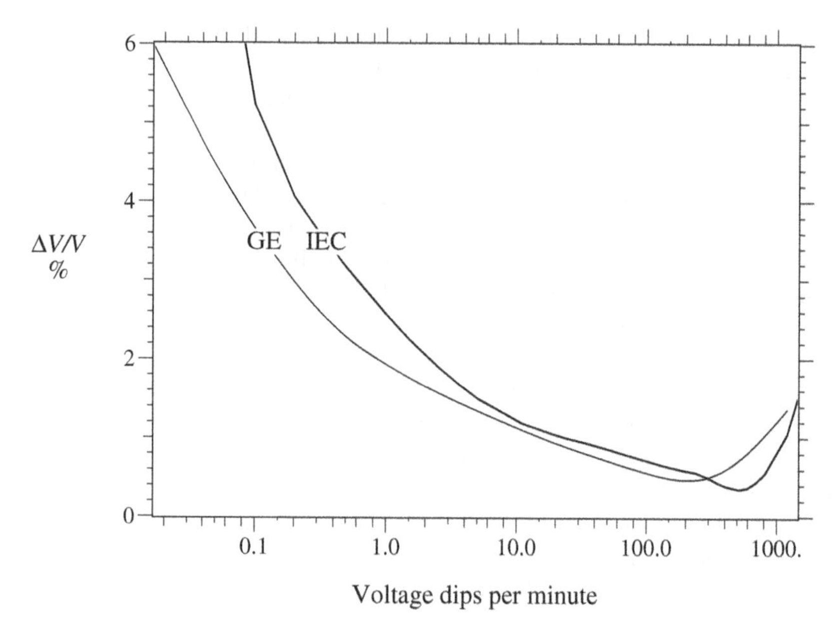

FIGURE 3.21
GE borderline of irritation curve compared to the voltage change necessary to produce $P_{st} = 1$ from the IEC flickermeter with a square-wave input.

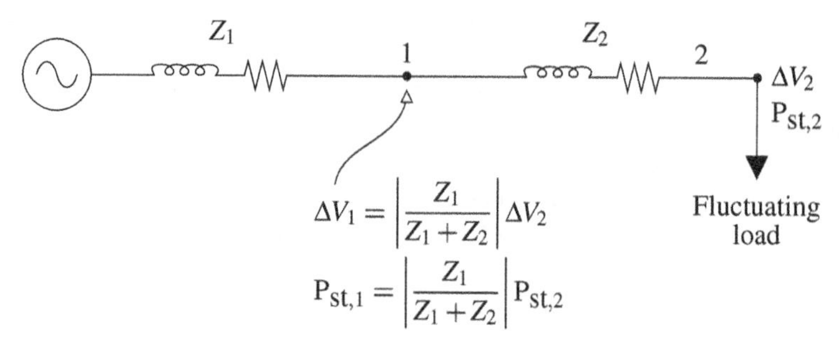

FIGURE 3.22
Propagation of voltage flicker.

$$P_{st,TOTAL} = \left(\sum P_{st,i}{}^{m}\right)^{\frac{1}{m}}$$

UIE, the organization that developed many of the European flicker standards (UIEPQ-9715, 1999), suggests the following values of m for combining multiple fluctuating loads (smaller m is more conservative):

$m = 4$: for arc furnaces operated to avoid simultaneous operations
$m = 3$: when the risk of simultaneous voltage variations is minimal (Most unrelated loads are in this category.)

$m = 2$: simultaneous operation of devices producing random-noise type fluctuations, such as arc furnaces operating at the same time
$m = 1$: when there is a high probability of simultaneous operation

The flickermeter can effectively quantify flicker for existing loads; it does not help much for predicting the flicker at new installations. It is possible to model the time variations of fluctuating loads and run a software "flicker-meter" on the results. If a similar load exists on the system, another approach is to measure P_{st} at the existing location and scale it by the relative differences between impedances at the existing location and the new location:

$$P_{st,B} = \frac{Z_B}{Z_A} P_{st,A}$$

where
$P_{st,A}$ = the short-term flicker indicator measured at a location with an impedance back to the source of Z_A (the three-phase short-circuit impedance)
$P_{st,\,B}$ = the short-term flicker indicator for the same fluctuating load moved to a point B which has a short-circuit impedance of Z_B

A basic screening criteria based on the stiffness of the distribution circuit and the amount of fluctuation also helps determine if flicker could be a problem at new installations. The Europeans have adopted a set of criteria depending on the rate of change of the load and the fluctuation relative to the stiffness of the supply (Table 3.5). The change in load (ΔI) is referenced to the short-circuit current at the nearest customer (I_{SC}). At less than ten changes per minute, with a short-circuit level on the primary of 1000 A, a load could fluctuate by 4 A (as measured on the primary) without causing flicker if the fluctuations are fewer than 10/min. North American limits could be less conservative than this because the lower-voltage North American light bulb is less sensitive to flicker than filaments at higher European voltages (120-V vs. 230-V bulbs).

Motor starts are a special case of voltage flicker. Most motors start only a few times per day, which is possibly off the flicker curve. While motors may

TABLE 3.5

Thresholds Where Fluctuating Loads Will Not Cause Flicker Problems per (IEC 61000-3-7, 1995; UIEPQ-9715, 1999)

Number of Voltage changes per minute, r	$\Delta I / I_{SC}$, %
$r > 200$	0.1
$10 \le r \le 200$	0.2
$r < 10$	0.4

TABLE 3.6

Voltage Flicker Limits at Several Utilities during Motor Starting

Utility	Voltage Criteria
Dense urban area	3%
Urban/suburban	3%
Suburban and rural	3%
Urban and rural	3%
Rural, mountainous	4 V (on a 120-V base)
Rural, mountainous	none

Source: Willis, H. L., *Power Distribution Planning Reference Book*, Marcel Dekker, New York, 1997.

TABLE 3.7

One Utility's Allowable Motor Starting Currents

System	Maximum Allowable Starting Current, A
Single Phase	
120 V	100
208 V	160
240 V	200
Three Phase	
208 V	1554
240 V	1346
480 V	673
2400 V	135

Note: Automatically controlled motors are limited to half of the allowable starting currents in the table.

start infrequently, when they do start, the voltage change is sharp, and light change can be deep and easily visible. Motors normally draw five to six times full-load current during starting. Normally, the voltage drops suddenly and then gradually recovers over several seconds as the motor comes up to speed.

Utilities normally have a criteria for motor starts based on the change of voltage and how often the customer starts the motor. Willis (1997) reported that many utilities use a criteria of 3% during motor starting (Table 3.6), but some utilities vary on how they define the percentage (either relative to the nominal voltage, the minimum voltage, or the voltage at the time of the motor start). Also, utilities often limit the size of motors or the starting current allowed by end users, depending on the voltage (see Table 3.7 for one utility's limits).

The source impedance at the customer with the motor (or other fluctuating load) is an important component. As with harmonics, we are most concerned

TABLE 3.8

Reference Impedances as Seen from the Electrical Panel Board

Location	Impedance, Ω
120-V residential, phase to neutral	$0.19 + j0.062$
240-V residential, phase to return phase	$0.20 + j0.080$
600-V industrial, phase to neutral	$0.58 + j0.107$
600-V industrial, phase to phase	$0.57 + j0.135$

Source: Ba, A. O., Bergeron, R., and Laperriere, A., "Source impedances of the Canadian distribution systems," CIRED 97, 1997.

about the voltage drop at the point of common coupling, the point where other customers can be tapped off. For customers fed from the same transformer, the point of common coupling is at the transformer. The impedance at the transformer is dominated by the transformer's impedance; the primary-side impedance is normally small. If customers share secondary circuits, also add the impedance of the shared secondary. For larger customers with their own dedicated transformer, the point of common coupling is on the primary. To evaluate the flicker to the customer that has the fluctuating load, consider the secondary and service drop to the meter. Hydro Quebec estimated reference impedances for Canadian residential and industrial facilities (Ba et al., 1997). At the panel, they estimated that 95% of customers have lower impedances than the values in Table 3.8, which we could use as a first approximation for motor starting and other fluctuation limits.

Interharmonics — harmonic distortions that are not integer multiples of the fundamental — can cause voltage flicker (Gunther, 2001). Noninteger harmonics are less common than integer harmonics, so this problem is not particularly widespread. Cycloconverters, arc furnaces, arc welders, and induction furnaces inject interharmonics. Interharmonics also come from loads like ovens or furnaces with integral cycle control, where the load controls the average voltage by either giving the heater full voltage or no voltage on a cycle-by-cycle basis (60% average voltage could come from six cycles on, then four cycles off). Standard six-pulse or twelve-pulse power converters do not normally create noninteger harmonics, but a converter can create noninteger harmonics if its electronic switches fire at the wrong time. Misfiring can come from incorrect control settings or a variety of hardware problems. The Wisconsin Electric Power Company had such a problem with a dc arc furnace that caused flicker (Tang et al., 1994).

Two superimposed frequencies beat against each other at this frequency:

$$f = |f_{ih} - f_0|$$

where

f_{ih} = frequency of the interharmonic

f_0 = integer multiple of the fundamental frequency closest to f_{ih}

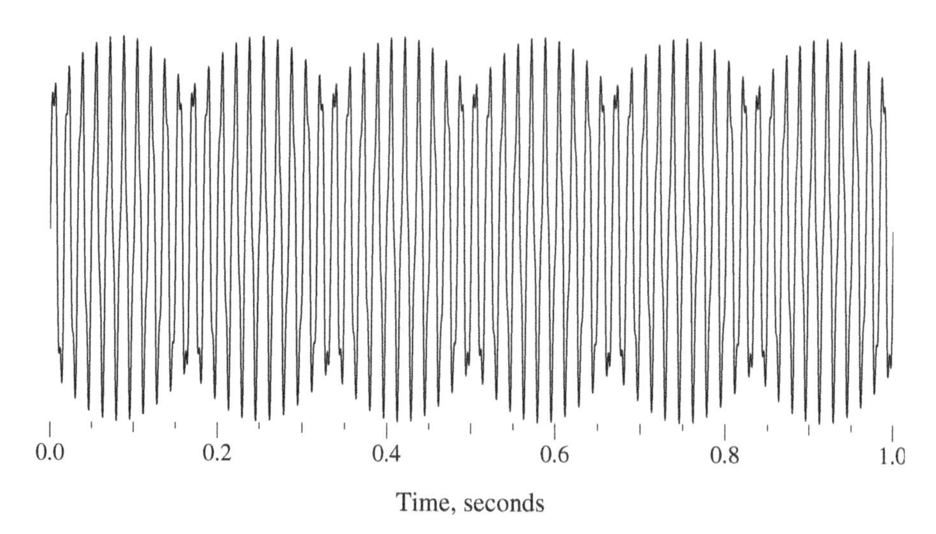

FIGURE 3.23
A 20% interharmonic frequency of $n = 3.1$ (186 Hz) causing a beat frequency of 6 Hz.

So, for $n = 3.1$, the beat frequency is $|3.1–3| = 0.1$ per unit, or $|186 \text{ Hz}–180 \text{ Hz}| = 6$ Hz (see Figure 3.23), right at the most sensitive flicker frequency. Frequencies of n = 1.9, 2.1, 2.9, 3.1, 3.9, and 4.1 per unit each beat at the same frequency: 0.1 per unit. Lower-frequency interharmonics — those near the first through third harmonic — are most likely to cause flicker. The IEC flickermeter detects flicker due to rms changes. Low-frequency interharmonics below the second harmonic change the rms, especially near the fundamental. But interharmonics above the second harmonic only modulate the peak; so at higher-frequency interharmonics, where the rms stays constant, the IEC flickermeter does not detect flicker (even though the fluctuations in the peak can result in noticeable flicker on fluorescent lights).

Incandescent lamps respond to rms voltage. Interharmonics below the second harmonic can cause flicker; higher frequencies do not. Fluorescent lights with electronic ballasts are particularly sensitive to the waveform peak; as the peak fluctuates, light output fluctuates. Electronic ballasts rectify the ac, which tracks the waveform peaks.

Just as capacitors can amplify integer harmonics, resonances can amplify noninteger harmonics. Solutions to flicker from interharmonics can include some of the solutions to regular flicker discussed in the next section. Also, harmonic solutions are appropriate in some situations, especially removing resonances by resizing or moving capacitors.

3.4.1 Flicker Solutions

3.4.1.1 Load Changes

Often, flicker is most economically solved at the source of the problem, the fluctuating load.

- *Welders* — For single large welders, changing the electrode firing sequence to draw smaller but more frequent current pulses can reduce flicker (EPRI PEAC Case Study No. 1, 1997). For multiple welders, control equipment can add a short delay to some welders to prevent simultaneous firing of several units. Also, on some welders, users can lower the weld "heat" (and current draw) without sacrificing the quality of the weld.
- *Motors* — For motor-starting flicker, one of several reduced-voltage starters will reduce the current draw and voltage deviation during starting. Common ways to reduce the voltage include an autotransformer start, a reactor start, and electronically switching the voltage (like a light dimmer) during starting. Reducing voltage to the motor during starting reduces the current inrush from 5 to 6 times normal current to under 2 per unit in some cases. Electronic motor starters can ramp the current to a preset maximum level. However, reduced-voltage starts also reduce the starting torque, so mechanical load issues can limit how much we can reduce the voltage. For motors susceptible to stalling (such as wood grinders), more careful operations on the mechanical side can reduce flicker (such as controlling the stock fed to the grinder).
- *Resistive heaters* — Heaters with on–off type controls flicker from repeated on and off cycling. Often, more precise control with less fluctuation is possible by upgrading the heater controls — regulating the voltage to the heating elements or splitting one heater into separately controlled elements.
- *Operational limits* — A simple option requiring no physical changes is to limit operation of offending loads to times when neighboring customers are unlikely to be bothered by flickering lights. Midnight to 6 a.m. is normally safe. Daytime operation might be tolerated. Operation from 6 p.m. to 11 p.m. is most likely to cause complaints. For facilities with multiple fluctuating loads, facilities can stagger operations to limit overlapping operation of multiple units.

Another option is to provide solutions where the customers are complaining. Review the conditions when flicker occurs and what lighting fixtures are flickering. A review of the lighting may reveal especially sensitive light fixtures such as some types of dimmers. These light fixtures can be replaced by fixtures that flicker less. One option is to convert from incandescent to fluorescent lights, which do not flicker as much.

If the facility causing the fluctuations is the only one experiencing the flicker, one solution is to isolate the lighting circuits from the circuits with fluctuating loads. Run separate circuits to the lighting loads. For single-phase fluctuating loads, put them on one phase and the lighting on the other two phases.

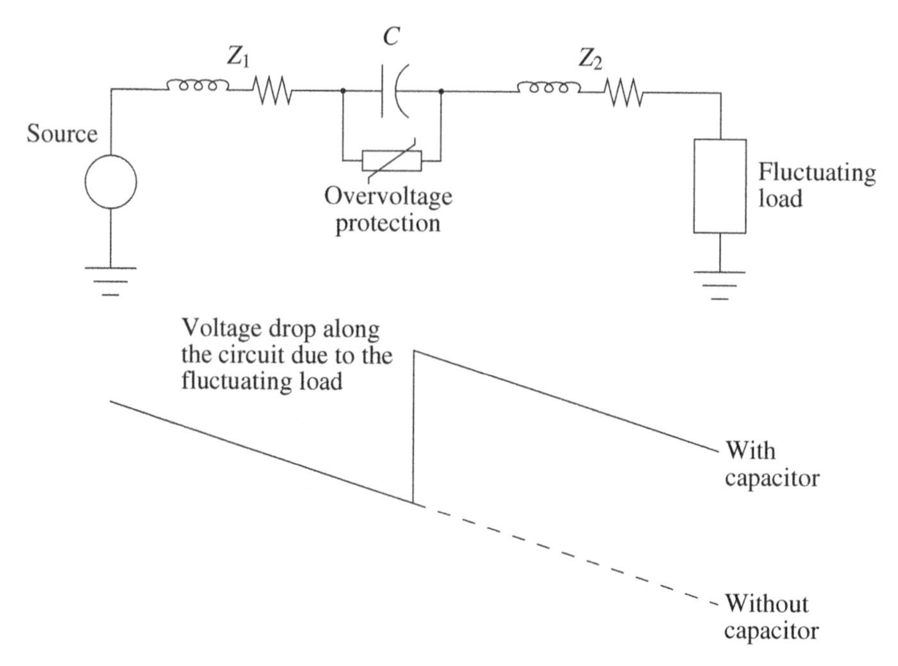

FIGURE 3.24
Series capacitor.

3.4.1.2 Series Capacitor

On distribution lines, much of the voltage drop is due to a circuit's inductance. If we add a capacitor in series with the inductance, the capacitor cancels out a portion of the inductance (see Figure 3.24). Now, the circuit has less total inductance, so fluctuations in load cause less voltage drop. A series capacitor is a passive circuit element. It responds instantaneously and automatically. Series capacitors help with voltage regulation as well as improving voltage flicker.

The best place for a series capacitor depends on where the complaining customers are relative to the offending load. Customers upstream of the series capacitor see no difference in flicker. Voltage improves downstream of the series capacitor. Sizing and placement involves tradeoffs. Ideally, we want to size the capacitor such that its impedance equals the circuit impedance to the fluctuating load (smaller series capacitor impedances are more expensive: more kvars, more units in parallel). For placement, a good rule of thumb is electrically half way between the source and the fluctuating load — the point where the voltage drop is half of the total voltage drop due to the fluctuating load. Locations closer to the source reduce flicker for more customers but increase fault duties. An IEEE working group suggests a rule-of-thumb location where the voltage drop is one-third to one-half of the total drop (Miske, 2001).

Series capacitors can superbly compensate flicker from fluctuating inductive loads. For resistive loads, series capacitors provide much less benefit. The voltage drop through the distribution system is approximately $I_R R + I_X X$. Reducing X with a series capacitor helps reduce the $I_X X$ term for fluctuating I_X; but if just I_R fluctuates, series capacitors provide little benefit.

Electrically, series capacitors elegantly solve voltage flicker; in practice, they are not widely used, mainly because of

- *Reliability of short-circuit protection* — Historically, spark gaps were used to protect the capacitors during downstream faults. Utilities have had many problems with failures.
- *Cost* — Series capacitors are nonstandard and must be custom engineered.
- *Unusual* — Line crews and field engineers find series capacitors strange. For example, if crews switch in a shunt capacitor downstream of a series capacitor, the voltage may go down instead of up. Increased fault currents downstream of the capacitor can make coordination of protective devices more difficult.
- *Ferroresonance* — Series capacitors may also ferroresonate with downstream transformers under the right conditions.

A key design issue is the voltage across the capacitor during faults downstream of the unit. During a fault downstream of the series capacitor, the voltage across the capacitor units is the short-circuit current times the capacitor impedance. Depending on the impedances upstream of the capacitor relative to the capacitor size, the voltage across the series capacitor can exceed the system's nominal voltage. When calculating the available fault current downstream of the capacitor, include the impedance reduction caused by the capacitor. The fault current normally can rise past the capacitor (depending on the amount of capacitance relative to the system impedance and the line X/R ratio). The highest fault current can be some distance from the capacitor. Figure 3.25 shows an example on a 12.5-kV system for a series capacitor placed 3 mi from the substation to compensate for a fluctuating load at 8 mi from the substation. The series capacitor significantly alters the fault-current profile, and the fault current impresses significant voltage across the capacitor. Some sort of protection is needed to prevent this.

The increased fault current raises the primary voltage and the customers' voltage during the fault. Figure 3.26 shows a profile of the line voltages for a fault at 4.75 mi for the example shown in Figure 3.25. Under this scenario, the fault current is leading (capacitive), so voltages rise along the line until the series capacitor. These overvoltages are unacceptable, so some overvoltage protection across the capacitor is vital to prevent them. This example has a rather poor choice of location and impedances, but it illustrates how important it is to consider these applications carefully. Reducing the impedance of the capacitor reduces the overvoltages and reduces the fault currents.

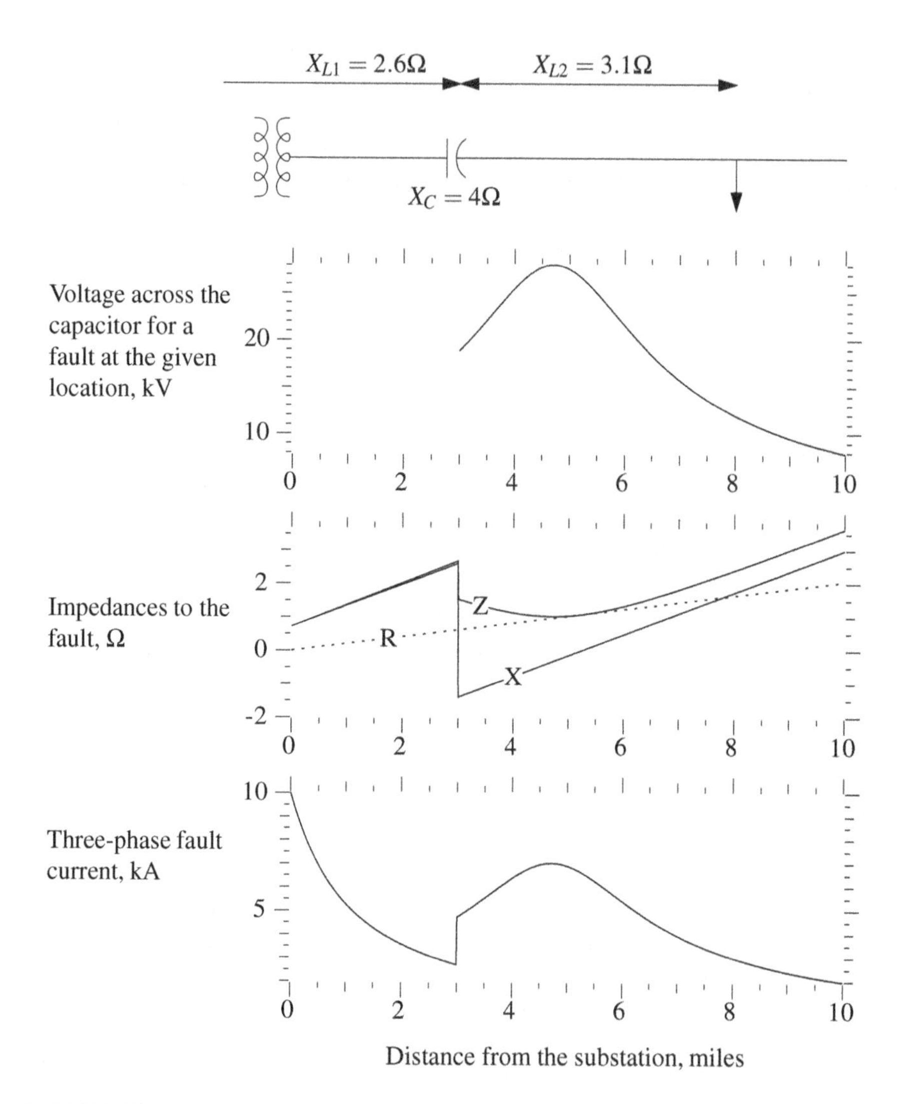

FIGURE 3.25
Fault-current profile on a 12.5-kV circuit with a series capacitor located 3 miles from the substation (the same line parameters as Figure 4.11, 500-kcmil AAC conductors). The series capacitor has no short-circuit protection.

Lowering the capacitor's impedance to maintain a total line impedance that is always reactive (never capacitive) eliminates the problem of overvoltages. But we still have higher fault currents downstream of the capacitor, and lowering the capacitor's impedance reduces the flicker-prevention capability.

Several protection arrangements have been used to protect capacitors from overvoltages. A spark gap is the simplest device but has been problematic. Duke Power reported that many of the gaps on their units kept arcing and started fires (Morgan et al., 1993). Gap erosion also reduced protection and

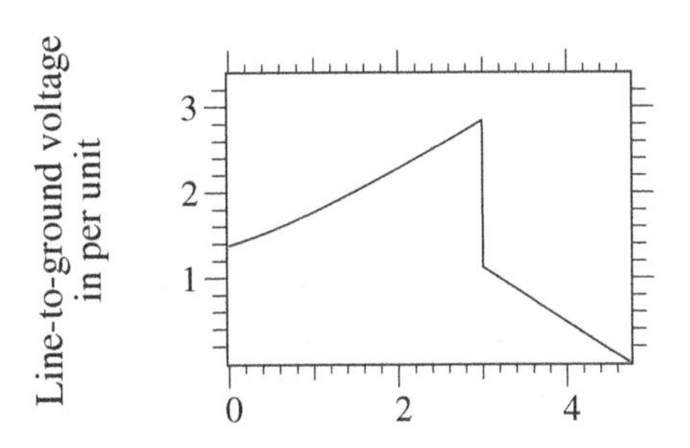

FIGURE 3.26
Voltages along the circuit for a three-phase fault at 4.75 miles for the application shown in Figure 3.25.

led to capacitor failures. Lat et al. (1990) used an improved gap that reduced erosion. The most modern units use metal-oxide arresters as protection against overvoltages; several stacks of metal-oxide blocks in parallel are needed to absorb the energy during a fault. Series capacitors normally have bypass switches to short the unit out if needed. Some designs close the bypass as soon as a fault is detected to reduce the duty on the capacitors and the arresters. Another benefit of using arresters for overvoltage protection is that the arresters short out the capacitors so that the series capacitor does not significantly increase the downstream fault current or cause overvoltages.

Marshall (1997) dealt with the problem of short-circuit protection by designing installations so that they did not need extra protection. This is possible at locations with low fault currents and where we can use a capacitor with a reasonably low voltage rating, especially if we do not try to compensate too aggressively. This is more likely on long rural circuits where the impedances are high, and we want a high capacitive impedance (which means we need less kvar in our series bank). Picking a higher voltage rating for a capacitor unit requires using several more units in parallel. For a given kvar rating, doubling the voltage rating raises the unit's impedance by a factor of four; now, we need four times the kvar for the same impedance.

3.4.1.3 Static Var Compensator

Static var compensators (SVCs) dynamically change their reactance, injecting or absorbing reactive power to regulate the voltage in real time. They are often used at large arc furnaces to control voltage flicker. Utilities sometimes use them to reduce voltage flicker on distribution circuits (Jatskevich et al., 1998; Kemerer and Berkebile, 1998; Wong et al., 1990).

Several styles of var compensators are available:

- *Switched inductor* — A thyristor controls how much reactive power is drawn from the system, depending on where on the voltage waveshape the thyristor turns on. Full vars are drawn if the thyristor turns on when the voltage is at its peak (a full half cycle of current is drawn). Less vars are drawn if the controller waits to turn on the thyristor. Once fired, the thyristor stays on until the current goes through zero and the thyristor shuts off. A drawback to this configuration is that the current blips are not sinusoidal and create harmonics. The inductor may be used in parallel with a shunt capacitor to control vars negatively and positively.

- *Switched capacitors* — In this configuration, several shunt capacitors each have thyristors to control whether the capacitor is connected. The number of capacitor units switched in determines the amount of vars injected. Having more individual blocks of capacitors helps keep the control smooth. The switched capacitors may be paralleled with a fixed inductor to allow the unit to control vars in both directions.
- *STATCOM* — The most advanced reactive power controller, this custom power device has a power electronics interface (normally a pulse-width-modulated inverter using IGBTs or other fast-switching electronics) that injects or absorbs reactive power that is temporarily stored in capacitors on a dc bus. The STATCOM (static compensator) can smoothly vary the reactive power with little harmonics. The STATCOM may be extended to inject or absorb real power with auxiliary energy storage.

None of these devices can instantly counter changes from fluctuating loads; some delay is always present. The delay is normally on the order of one-half to two cycles, depending on the compensator and controller technology.

Reactive power is more efficient than real power at countering voltage changes. Reactive power is more efficient because the equivalent source impedance is normally more reactive than resistive, and the voltage drop is approximately

$$V_{drop} \approx R \cdot I_R + X \cdot I_X$$

where R and X are the source resistance and reactance, I_R is the resistive component of the load, and I_X is the reactive component of the load.

3.4.1.4 Other Solutions

Other utility-side solutions to flicker include

- *Dedicated circuit* — Running a separate load to the customer with a fluctuating load isolates the flicker to the customer causing the flicker. We can tap the dedicated circuit where the voltage change due to the fluctuating load is tolerable.
- *Reconductoring* — Larger conductors have lower resistance and slightly lower reactance. If the voltage fluctuation is due to changes

in the resistive load, larger conductors will help; but if the voltage fluctuation is due to reactive load, larger conductors will not help. Tighter phase spacings reduce the reactance somewhat, but not dramatically. Converting to aerial or underground cable does dramatically reduce the reactance.

- *Upgrading voltage* — On a percentage basis, voltage drop decreases as the square of system voltage. Of course, upgrading the system voltage requires new insulators, arresters, switches, and transformers.

Unfortunately, all of these are normally expensive options. Jenner and Brockhurst (1990) reported success using vacuum-switched shunt capacitor banks to counteract the voltage drop from large motor starts, although the measurements shown still left a significant voltage change.

3.5 Voltage Unbalance

Three-phase end-use equipment expects balanced voltages, each with the same magnitude. If the voltages are not balanced, some equipment can perform poorly, mainly induction motors. Voltage unbalance (or imbalance, an equivalent term used by many) is defined as:

$$\text{Percent Unbalance} = \frac{\text{Maximum deviation from the average}}{\text{Average of the three phase - to - phase voltages}} \times 100\%$$

Normally, phase-to-phase voltages are used for voltage unbalance rather than phase-to-ground voltages.

ANSI C84.1-1995 states that utilities should limit the maximum voltage unbalance to 3%, measured under no-load conditions. Most distribution systems have modest unbalance, well under the ANSI 3% limit. Field surveys sited by ANSI C84.1 report that only 2% of electric supply systems have unbalance that exceeds 3%.

Motors are most efficient with balanced voltages; unbalanced voltage heats motors significantly due to negative-sequence currents. When applied to unbalanced voltages, motors should be derated. Figure 3.27 shows a derating factor that ANSI and NEMA provides for motors.

Motors have relatively low impedance to negative-sequence voltage (approximately equal to the motor's locked rotor impedance); therefore, a small negative-sequence component of the voltage produces a relatively large negative-sequence current. The negative-sequence voltage creates a rotating field that rotates in the opposite direction that the motor is spinning. The rotor is a shorted winding to this counter-rotating field.

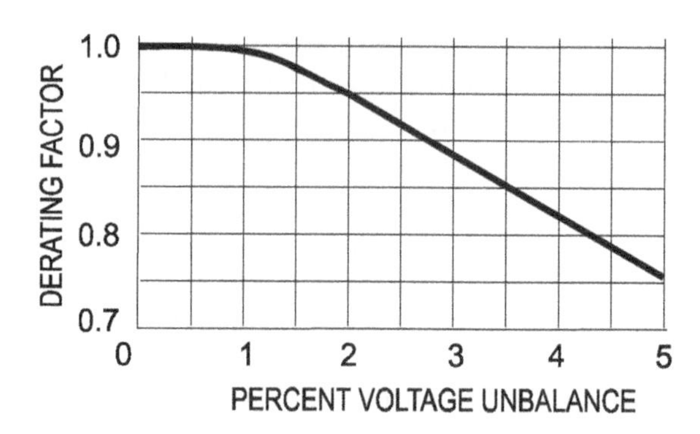

FIGURE 3.27
Motor derating curve based on voltage unbalance. (From NEMA MG-1, *Standard for Motors and Generators*, 1998. With permission.)

NEMA standard MG-1 allows a 1% voltage unbalance without derating; operation above 5% unbalance is not recommended (NEMA MG-1, 1998).

Because motors are really influenced by the negative-sequence voltage, another more-precise indicator is the unbalance factor, defined in European standards:

$$\%VUF = \frac{V_2}{V_1} \times 100\%$$

where
V_1 = positive-sequence voltage
V_2 = negative-sequence voltage

The percentage negative-sequence voltage is close to unbalance as defined by NEMA and ANSI using phase-to-phase voltage measurements, with a difference of at most 13% between the two terms (UIEPQ-9715, 1996). The EPRI DPQ study found that 98% of the time, including all sites monitored, the unbalance factor (V_2/V_1) was less than 1.5% with an average just above 0.8% (EPRI TR-106294-V2, 1996). In addition, only 2 out of 277 sites averaged above a 3% unbalance factor.

While negative-sequence voltages impact motors and other line-to-line connected loads the most, zero-sequence unbalance can affect line-to-ground connected three-phase loads. Most sites have little zero-sequence voltage unbalance (although the current unbalance may be significant). The DPQ study found similar levels of V_0/V_1 as V_2/V_1: V_0/V_1 averaged just over 0.8%; only 3 of 277 sites averaged above 3% zero-sequence voltage. Whereas transformer connections do not change the ratio of V_2/V_1, zero-sequence voltages are blocked by ungrounded transformer connections like the delta – grounded wye.

While motors receive much of the attention, unbalance also affects other loads, particularly electronic loads like adjustable-speed drives. With voltage unbalance, adjustable-speed drives draw significantly unbalanced currents, currents so unbalanced that overload protection circuits can trip (EPRI PEAC Brief No. 28, 1995). An adjustable-speed drive rectifies the three incoming phases. Each phase-to-phase input takes its turn charging up the capacitor on the drive's dc bus; but if one of these three inputs has higher voltage than the others, that phase conducts significantly more current than the other two. Even a 3% voltage unbalance can cause some drives to trip. Voltage unbalance also increases the harmonics that drives create, including the third harmonic, which balanced three-phase drives do not normally produce. Voltage unbalance also increases the ripple on the dc bus voltage.

Some causes of voltage unbalance in the order of most to least likely include:

- *Current unbalance* — Voltage unbalance normally comes from excessive current unbalance — more heavily loaded phases have more voltage drop than the lightly loaded phases. Large single-phase laterals are often culprits. Normally, crews can change the phase connections of single-phase laterals to balance the voltages. Also consider the possibility of shifting load to another circuit.
- *Capacitor banks with blown fuses* — Banks with blown fuses inject unbalanced vars, which can severely unbalance voltages.
- *Malfunctioning regulators* — Single-phase line voltage regulators adjust each phase independently. If one phase is stuck or is otherwise working improperly or has inappropriate settings, the regulator will unbalance the voltages. Note that because regulators adjust each phase independently, they perform well for improving a circuit's voltage balance.
- *Open-wye–open-delta transformers* — These banks can supply three-phase load from a two-phase supply. But the connection is prone to causing voltage unbalance to the end user on the transformer. Also, because it injects significant current into the ground, it can contribute to voltage unbalance elsewhere on the circuit.
- *Untransposed circuit* — Transposing distribution circuits is almost unheard of; it is just not needed. Rarely, balanced load can cause unbalanced voltages because the line impedances are unbalanced. As an example, on a flat configuration with an 8-ft crossarm with 336-kcmil ACSR conductors at 12.47 kV, a balanced, unity-power factor current of 100 A through 10 mi of line causes 1% voltage unbalance at the end.

Normally, we can solve unbalanced voltages by correcting the current balance or fixing a problem regulator or capacitor. After that, adding voltage regulators is an option.

References

Abetti, P. A., Johnson, I. B., and Schultz, A. J., "Surge Phenomena in Large Unit-Connected Steam Turbine Generators," *AIEE Transactions*, vol. 71, part III, pp. 1035–47, December 1952.

Ba, A. O., Bergeron, R., and Laperriere, A., "Source impedances of the Canadian distribution systems," CIRED 97, 1997.

Bellei, T. A., O'Leary, R. P., and Camm, E. H., "Evaluating Capacitor-Switching Devices for Preventing Nuisance Tripping of Adjustable-Speed Drives Due to Voltage Magnification," *IEEE Transactions on Power Delivery*, vol. 11, no. 3, pp. 1373–8, July 1996.

Bohn, J., "Utility Concerns for Electric Vehicle Battery Charger Penetration," Panel on Utility Application of Harmonic Limits, IEEE PES Summer Power Meeting, 1996.

Dwyer, R., Khan, A. K., McGranaghan, M., Tang, L., McCluskey, R. K., Sung, R., and Houy, T., "Evaluation of Harmonic Impacts from Compact Fluorescent Lights on Distribution Systems," IEEE/PES Winter Power Meeting, 1995.

Emanuel, A. E., Janczak, J., Pileggi, D. J., Gulachenski, E. M., Breen, M., Gentile, T. J., and Sorensen, D., "Distribution Feeders with Nonlinear Loads in the North-east USA. I. Voltage Distortion Forecast," *IEEE Transactions on Power Delivery*, vol. 10, no. 1, pp. 340–7, January 1995.

EPRI PEAC Application #6, *Avoiding Harmonic-Related Overload of Shared-Neutral Conductors*, 1996.

EPRI PEAC Brief No. 21, *The Effects of Neutral-to-Ground Voltage Differences on PC Performance*, EPRI PEAC, Knoxville, TN, 1994.

EPRI PEAC Brief No. 25, *The Effect of Lamp Dimmers on Light Flicker*, EPRI PEAC, Knoxville, TN, 1995.

EPRI PEAC Brief No. 28, *Input Performance of ASDs During Supply Voltage Unbalance*, EPRI PEAC, Knoxville, TN, 1995.

EPRI PEAC Brief No. 36, *Lamp Flicker Predicted by Gain-Factor Measurements*, EPRI PEAC, Knoxville, TN, 1996.

EPRI PEAC Case Study No. 1, *Light Flicker Caused By Resistive Spot Welder*, EPRI PEAC, Knoxville, TN, 1997.

EPRI PEAC Solution No. 1, *Equalizing Potential Differences*, EPRI PEAC, Knoxville, TN, 1993.

EPRI TR-106294-V2, *An Assessment of Distribution System Power Quality. Volume 2: Statistical Summary Report*, Electric Power Research Institute, Palo Alto, CA, 1996.

Etezadi-Amoli, M. and Florence, T., "Voltage and Current Harmonic Content of a Utility System — A Summary of 1120 Test Measurements," *IEEE Transactions on Power Delivery*, vol. 5, no. 3, pp. 1552–7, July 1990.

General Electric GET-1008L, "Distribution Data Book."

Gorgette, F. A., Lachaume, J., and Grady, W. M., "Statistical Summation of the Harmonic Currents Produced by a Large Number of Single Phase Variable Speed Air Conditioners: a Study of Three Specific Designs," *IEEE Transactions on Power Delivery*, vol. 15, no. 3, pp. 953–9, July 2000.

Govindarajan, S. N., Cox, M. D., and Berry, F. C., "Survey of Harmonic Levels on the Southwestern Electric Power Company System," *IEEE Transactions on Power Delivery*, vol. 6, no. 4, pp. 1869–75, October 1991.

Greenwood, A., *Electrical Transients in Power Systems*, 2nd ed., Wiley Interscience, New York, 1991.

Gunther, E. W., "Interharmonics in Power Systems," IEEE Power Engineering Society Summer Meeting, 2001.

Halpin, S. M., Bergeron, R., Blooming, T., Burch, R. F., Conrad, L. E., and Key, T., "Voltage and Lamp Flicker Issues: Should the IEEE Adopt the IEC Approach?" IEEE P1453 working group paper. Downloaded from http://www.manta.ieee.org/groups/1453/drpaper.html.

IEC 868, *Flickermeter: Functional and Design Specifications*, International Electrical Commission (IEC), 1986.

IEC 61000-3-7, *Limitation of Voltage Fluctuation and Flicker for Equipments Connected to Medium and High Voltage Power Supply Systems*, International Electrical Commission (IEC), 1995.

IEEE P519A draft 7, *Guide for Applying Harmonic Limits on Power Systems*, 2000.

IEEE Std. 141-1993, *IEEE Recommended Practice for Electric Power Distribution for Industrial Plants.*

IEEE Std. 142-1991, *IEEE Recommended Practice for Grounding of Industrial and Commercial Power Systems.*

IEEE Std. 519-1992, *IEEE Recommended Practices and Requirements for Harmonic Control in Electrical Power Systems.*

IEEE Std. 776-1992, *IEEE Recommended Practice for Inductive Coordination of Electric Supply and Communication Lines.*

IEEE Std. 1100-1999, *IEEE Recommended Practice for Powering and Grounding Electronic Equipment.*

IEEE Std. 1137-1991, *IEEE Guide for the Implementation of Inductive Coordination Mitigation Techniques and Application.*

IEEE Std. C57.110-1998, *IEEE Recommended Practice for Establishing Transformer Capability When Supplying Nonsinusoidal Load Currents.*

IEEE Task Force, "The Effects of Power System Harmonics on Power System Equipment and Loads," *IEEE Transactions on Power Apparatus and Systems*, vol. PAS-104, no. 9, pp. 2555–63, September, 1985.

IEEE Task Force, "Effects of Harmonics on Equipment," *IEEE Transactions on Power Delivery*, vol. 8, no. 2, pp. 672–80, April 1993.

IEEE Working Group, "Power Line Harmonic Effects on Communication Line Interference," *IEEE Transactions on Power Apparatus and Systems*, vol. PAS-104, no. 9, pp. 2578–87, September 1985.

Jatskevich, J., Wasynczuk, O., and Conrad, L., "A Method of Evaluating Flicker and Flicker-Reduction Strategies in Power Systems," *IEEE Transactions on Power Delivery*, vol. 13, no. 4, pp. 1481–7, October 1998.

Jenner, M. and Brockhurst, F. C., "Vacuum Switched Capacitors to Reduce Induction Motor Caused Voltage Flicker on a 12.47 kV Rural Distribution Line," IEEE Rural Electric Power Conference, 1990.

Kemerer, R. S. and Berkebile, L. E., "Directly Connected Static VAr Compensation in Distribution System Applications," IEEE Rural Electric Power Conference, 1998.

Kolcio, N., Halladay, J. A., Allen, G. D., and Fromholtz, E. N., "Transient Overvoltages and Overcurrents on 12.47 kV Distribution Lines: Field Test Results," *IEEE Transactions on Power Delivery*, vol. 7, no. 3, pp. 1359–70, July 1992.

Lat, M. V., Kundu, D., and Bonadie, G., "Overvoltage Protection Scheme for Series Capacitor Banks on High Voltage Distribution Systems," *IEEE Transactions on Power Delivery,* vol. 5, no. 3, pp. 1459–65, July 1990.

Mansoor, A., Grady, W. M., Staats, P. T., Thallam, R. S., Doyle, M. T., and Samotyj, M. J., "Predicting the Net Harmonic Currents Produced by Large Numbers of Distributed Single-Phase Computer Loads," *IEEE Transactions on Power Delivery,* vol. 10, no. 4, pp. 2001–6, October 1995.

Marshall, M. W., "Using Series Capacitors to Mitigate Voltage Flicker Problems," IEEE Rural Electric Power Conference, 1997.

McGranaghan, M. F., Grebe, T. E., Hensley, G., Singh, T., and Samotyj, M., "Impact of Utility Switched Capacitors on Customer Systems. II. Adjustable-Speed Drive Concerns," *IEEE Transactions on Power Delivery,* vol. 6, no. 4, pp. 1623–8, October 1991.

McGranaghan, M. F., Zavadil, R. M., Hensley, G., Singh, T., and Samotyj, M., "Impact of Utility Switched Capacitors on Customer Systems — Magnification at Low Voltage Capacitors," *IEEE Transactions on Power Delivery,* vol. 7, no. 2, pp. 862–8, April 1992.

Miske, S. A., "Considerations for the Application of Series Capacitors to Radial Power Distribution Circuits," *IEEE Transactions on Power Delivery,* vol. 16, no. 2, pp. 306–18, April 2001.

Morgan, L., Barcus, J. M., and Ihara, S., "Distribution Series Capacitor with High-Energy Varistor Protection," *IEEE Transactions on Power Delivery,* vol. 8, no. 3, pp. 1413–19, July 1993.

Nejdawi, I. M., Emanuel, A. E., Pileggi, D. J., Corridori, M. J., and Archambeault, R. D., "Harmonics Trend in NE USA: A Preliminary Survey," *IEEE Transactions on Power Delivery,* vol. 14, no. 4, pp. 1488–94, October 1999.

NEMA MG-1, *Standard for Motors and Generators,* 1998.

Pileggi, D. J., Gulachenski, E. M., Root, C. E., Gentile, T. J., and Emanuel, A. E., "The Effect of Modern Compact Fluorescent Lights on Voltage Distortion," *IEEE Transactions on Power Delivery,* vol. 8, no. 3, pp. 1451–9, July 1993.

Rice, D. E., "Adjustable Speed Drive and Power Rectifier Harmonics — Their Effect on Power Systems Components," *IEEE Transactions on Industry Applications,* vol. IA-22, no. 1, pp. 161–77, January/February 1986.

Sabin, D. D., Brooks, D. L., and Sundaram, A., "Indices for Assessing Harmonic Distortion from Power Quality Measurements: Definitions and Benchmark Data," *IEEE Transactions on Power Delivery,* vol. 14, no. 2, pp. 489–96, April 1999a.

Sabin, D. D., Grebe, T. E., and Sundaram, A., "Assessing Distribution System Transient Overvoltages Due to Capacitor Switching," International Conference on Power Systems Transients, Technical University of Budapest, Budapest, Hungary, 1999b.

Schultz, A. J., Johnson, J. B., and Schultz, N. R., "Magnification of Switching Surges," *AIEE Transactions on Power Apparatus and Systems,* pp. 1418–26, February 1959.

Seebald, R. C., Buch, J. F., and Ward, D. J., "Flicker Limitations of Electric Utilities," *IEEE Transactions on Power Apparatus and Systems,* vol. PAS-104, no. 9, pp. 2627–31, September 1985.

Shankland, L. A., Feltes, J. W., and Burke, J. J., "The Effect of Switching Surges on 34.5 kV System Design and Equipment," *IEEE Transactions on Power Delivery,* vol. 5, no. 2, pp. 1106–12, April 1990.

Tang, L., Mueller, D., Hall, D., Samotyj, M., and Randolph, J., "Analysis of DC Arc Furnace Operation and Flicker Caused by 187 Hz Voltage Distortion," *IEEE Transactions on Power Delivery*, vol. 9, no. 2, pp. 1098–107, April 1994.

Thallam, R. S., Grady, W. M., and Samotyj, M. J., "Estimating Future Harmonic Distortion Levels In Distribution Systems Due To Single-Phase Adjustable-Speed-Drive Air Conditioners: A Case Study," ICHPS V International Conference on Harmonics in Power Systems, 1992.

Troedsson, C. G., Gramlich, E. F., Gustin, R. F., and McGranaghan, M. F., "Magnification of Switching Surges as a Result of Capacitor Switching on a 34.5 kV Distribution System," Proceedings of the American Power Conference, 1983.

UIEPQ-9715, *Guide to Quality of Electrical Supply for Industrial Installation — Part 4: Voltage Unbalance*, Union Internationale de Electrothermie (UIE), 1996.

UIEPQ-9715, *Guide to Quality of Electrical Supply in Industrial Installations — Part 5: Flicker and Voltage Fluctuations*, Union Internationale de Electrothermie (UIE), 1999.

Walker, M. K., "Electric Utility Flicker Limitations," *IEEE Transactions on Industry Applications*, vol. IA-15, no. 6, pp. 644–55, November/December 1979.

Walling, R. A., Melchior, R. D., and McDermott, B. A., "Measurement of Cable Switching Transients in Underground Distribution Systems," *IEEE Transactions on Power Delivery*, vol. 10, no. 1, pp. 534–9, January 1995.

Willis, H. L., *Power Distribution Planning Reference Book*, Marcel Dekker, New York, 1997.

Wong, W. K., Osborn, D. L., and McAvoy, J. L., "Application of Compact Static VAR Compensators to Distribution Systems," *IEEE Transactions on Power Delivery*, vol. 5, no. 2, pp. 1113–20, April 1990.

Although they try hard, bless their hearts, consumers that diagnose their own powerline problems tend to miss a thing or two at times.

Powerlineman law #18, By CD Thayer and other Power Linemen, http://www.cdthayer.com/lineman.htm

4

Faults

Faults kill. Faults start fires. Faults force interruptions. Faults create voltage sags. Tree trimming, surge arresters, animal guards, cable replacements: these tools reduce faults. We cannot eliminate all faults, but appropriate standards and maintenance practices help in the battle. When faults occur, we have ways to reduce their impacts. This chapter focuses on the general characteristics of faults and specific analysis of common fault types with suggestions on how to reduce them. One of the definitions of a fault is (ANSI/IEEE Std. 100-1992):

Fault: A physical condition that causes a device, a component, or an element to fail to perform in a required manner; for example, a short circuit or a broken wire.

A fault almost always involves a short circuit between energized phase conductors or between a phase and ground. A fault may be a bolted connection or may have some impedance in the fault connection. The term "fault" is often used synonymously with the term "short circuit" defined as (ANSI/IEEE Std. 100-1992):

Short circuit: An abnormal connection (including an arc) of relatively low impedance, whether made accidentally or intentionally, between two points of different potential. (*Note:* The term *fault* or *short-circuit fault* is used to describe a short circuit.)

When a short-circuit fault occurs, the fault path explodes in an intense arc. Local customers endure an interruption, and customers farther away, a voltage sag; faults cause most reliability and power quality problems. Faults kill and injure line operators. Crew operating practices, equipment, and training must account for where fault arcs are likely to occur and must minimize crew exposure.

4.1 General Fault Characteristics

There are many causes of faults on distribution circuits. A large EPRI study was done to characterize distribution faults in the 1980s at 13 utilities mon-

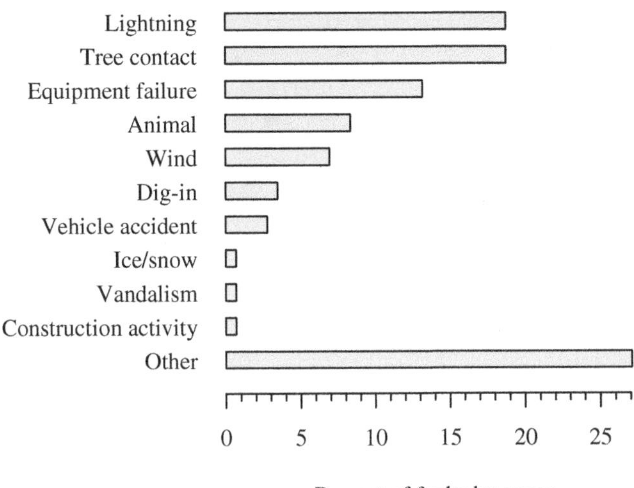

FIGURE 4.1
Fault causes measured in the EPRI fault study. (Data from [Burke and Lawrence, 1984; EPRI 1209-1, 1983].)

itoring 50 feeders (Burke and Lawrence, 1984; EPRI 1209-1, 1983). The distribution of permanent fault causes found in the EPRI study is shown in Figure 4.1. Many of the fault causes are discussed in more detail in this chapter. Approximately 40% of faults in this study occurred during periods of adverse weather which included rain, snow and ice.

Distribution faults occur on one phase, on two phases, or on all three phases. Single-phase faults are the most common. Almost 80% of the faults measured involved only one phase either in contact with the neutral or with ground (see Table 4.1). As another data point, measurements on 34.5-kV

TABLE 4.1

Number of Phases Involved in Each Fault Measured in the EPRI Fault Study

Fault	Percentage
One phase to neutral	63%
Phase to phase	11%
Two phases to neutral	2%
Three phase	2%
One phase on the ground	15%
Two phases on the ground	2%
Three phases on the ground	1%
Other	4%

Source: Burke, J. J. and Lawrence, D. J., "Characteristics of Fault Currents on Distribution Systems," *IEEE Transactions on Power Apparatus and Systems*, vol. PAS-103, no. 1, pp. 1–6, January 1984. EPRI 1209-1 (1983).

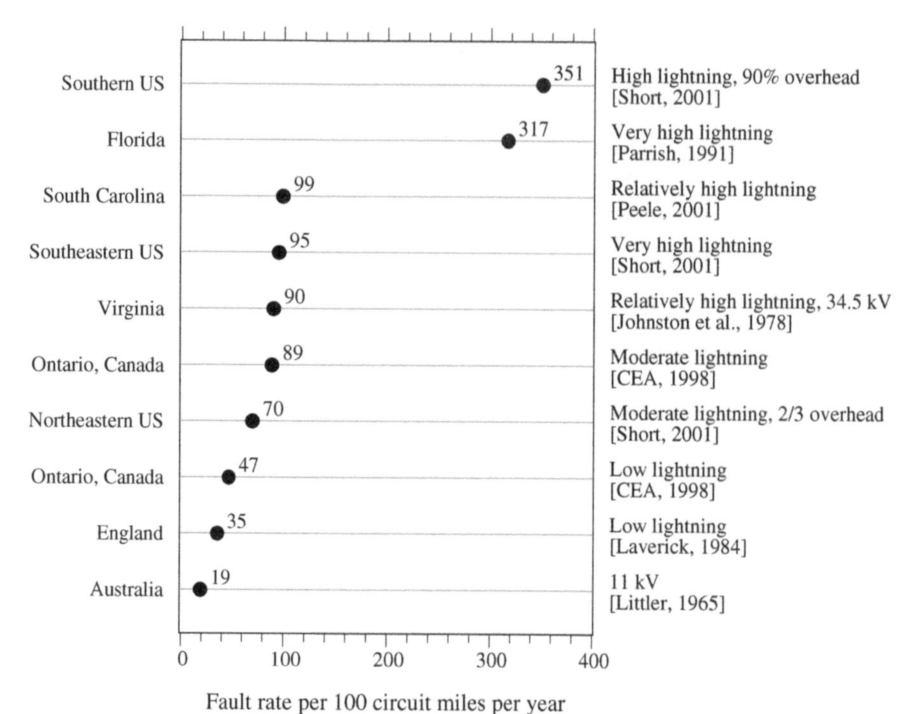

FIGURE 4.2
Fault rates found in different studies. (Data from [CEA 160 D 597, 1998; Johnston et al., 1978; Laverick, 1984; Littler, 1965; Parrish, 1991; Peele, 2001; Short, 2001].)

feeders found that 75% of faults involved ground (also 54% were phase to ground, and 15% were phase to phase) (Johnston et al., 1978). Most faults are single phase because most of the overall length of distribution lines is single phase, so any fault on single-phase sections would only involve one phase. Also, on three-phase sections, many types of faults tend to occur from phase to ground. Equipment faults and animal faults tend to cause line-to-ground faults. Trees can also cause line-to-ground faults on three-phase structures, but line-to-line faults are more common. Lightning faults tend to be two or three phases to ground on three-phase structures.

Figure 4.2 shows fault rates found in various studies for predominantly overhead circuits. Ninety faults per 100 mi per year (55 faults/100 km/year) is common for utilities with moderate lightning. Fault rates increase significantly in higher lightning areas. This type of data is difficult to obtain. Utilities more commonly track faults that cause sustained interruptions, interruptions that contribute to reliability indices such as SAIDI (some data on these faults is shown in Figure 4.3). The actual fault rates are higher than this because many temporary faults are cleared by reclosing circuit breakers or reclosers.

Faults are either *temporary* or *permanent*. A permanent fault is one where permanent damage is done to the system. This includes insulator failures,

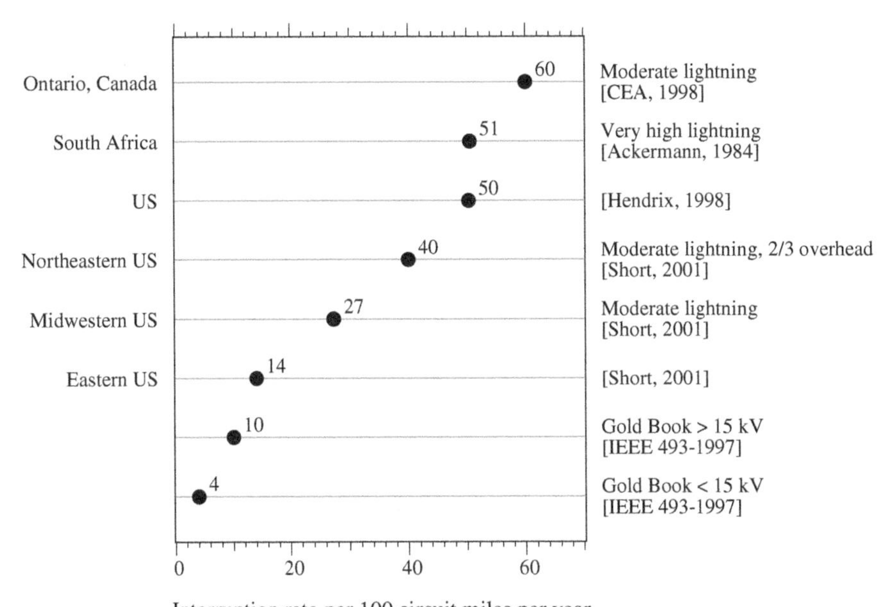

FIGURE 4.3
Fault rates for faults that cause sustained interruptions. (Data from [Ackermann and Baker, 1984; CEA 160 D 597, 1998; Hendrix, 1998; IEEE Std. 493-1997; Johnston et al., 1978; Parrish, 1991; Short, 2001].)

broken wires, or failed equipment such as transformers or capacitors. Virtually all faults on underground equipment are permanent. Most equipment fails to a short circuit. Permanent faults on distribution circuits usually cause sustained interruptions for some customers. To clear the fault, a fuse, recloser, or circuit breaker must operate to interrupt the circuit. The most critical location is the three-phase mains, since a fault on the main feeder will cause an interruption to all customers on the circuit. A permanent fault also causes a voltage sag to customers on the feeder and on adjacent feeders. Permanent faults may cause momentary interruptions for a customer. A common example is a fault on a fused lateral (tap). With *fuse saving* (where an upstream circuit breaker or recloser attempts to open before the tap fuse blows), a permanent fault causes a momentary interruption for customers downstream of the circuit breaker or recloser. After the first attempt to save the fuse, if the fault is still there, the circuit breaker allows the fuse to clear the fault. If a fault is permanent, all customers on the circuit experience a momentary — and the customers on the fused lateral experience a sustained interruption.

A temporary fault does not permanently damage any system equipment. If the circuit is interrupted and then reclosed after a delay, the system operates normally. Temporary (non-damage) faults make up 50 to 90% of faults on overhead distribution systems. The causes of temporary faults include lightning, conductors slapping together in the wind, tree branches that fall across conductors and then fall or burn off, animals that cause faults and

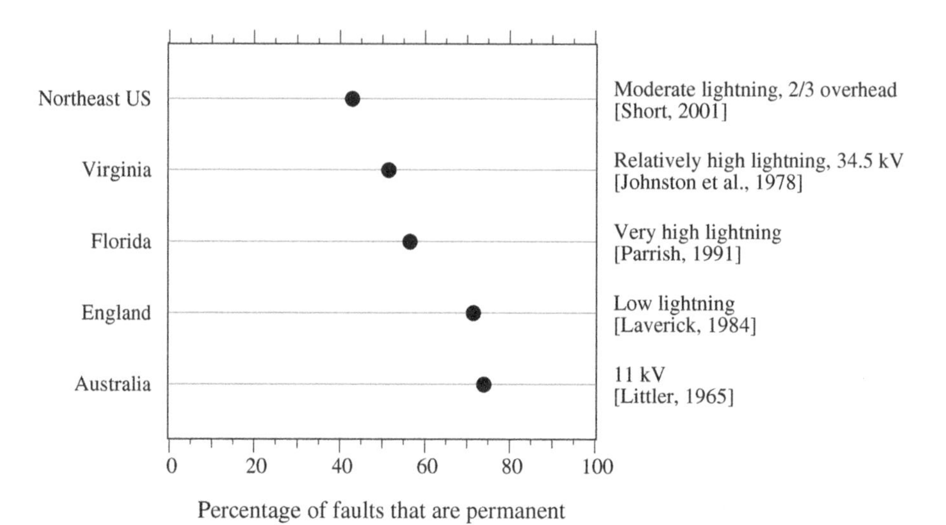

FIGURE 4.4
Percentage of faults that are permanent (on predominantly overhead circuits). (Data from [Johnston et al., 1978; Laverick, 1984; Littler, 1965; Parrish, 1991; Short, 2001].)

fall off, and insulator flashovers caused by pollution. Temporary faults are the main reason that reclosing is used almost universally on distribution circuit breakers and reclosers (on overhead circuits). Temporary faults will cause voltage sags for customers on the circuit with the fault and possibly for customers on adjacent feeders. Temporary faults cause sustained interruptions if the fault is downstream of a fuse, and fuse saving is not used or is not successful. For temporary faults on the feeder backbone, all customers on the circuit are momentarily interrupted. Faults that are normally temporary can turn into permanent faults. If the fault is allowed to remain too long, the fault arc can do permanent damage to conductors, insulators, or other hardware. In addition, the fault current flowing through equipment can do damage. The most common damage of this type is to connectors or circuit interrupters such as fuses.

The majority of faults on overhead distribution circuits are temporary. The data in Figure 4.4 confirms the widely held belief that 50 to 80% of faults are temporary. This very limited data set shows high lightning areas with lower percentages of temporary faults. This contradicts the notion that temporary faults are higher in areas with more lightning. Storms with lightning and wind should cause more temporary faults.

Determining the percentage of faults that are temporary versus permanent is complicated. For faults that operate fuses, it is easy. If it can be successfully re-fused without any repair, the fault is temporary:

$$\text{Percent permanent} = \frac{\text{Fuses replaced after repair}}{\text{Total number of fuse operations}} \times 100\%$$

Circuit breaker or recloser operations are difficult. If *fuse blowing* is used, where tap fuses always operate before the circuit breaker or recloser, the percentage of temporary faults cleared by circuit breakers and reclosers is:

$$\text{Percent permanent} = \frac{\text{Number of lockouts}}{\text{Number of lockouts} + \text{Number of successful reclose sequences}} \times 100\%$$

A SCADA system produces these numbers, but if this information is not available, the percentage can be approximated using circuit breaker count numbers:

$$\text{Percent permanent} = \frac{l}{n\text{-}r \cdot l} \times 100\%$$

where

n = total number of circuit breaker (or recloser) operations

r = number of reclose attempts before lockout (there are r+1 circuit breaker operations during a lockout cycle)

l = number of lockouts

If *fuse saving* is used, where the circuit breaker operates before lateral fuses, then it is more difficult to estimate the number of temporary faults. For the whole circuit (it is not possible to separate the faults on the mains from the faults on the taps), we can estimate the percentage as follows:

$$\text{Percent permanent} = \frac{l + f}{l + s + f_2} \times 100\%$$

where

s = number of successful reclose sequences

f = number of fuses replaced following repair (not including nuisance fuse operations)

f_2 = number of fuse operations that are not coincident with circuit breaker trips

f_2 should be close to zero, since the circuit breaker should operate for all faults. Assuming f_2 is zero (which may have to be done, since this is a difficult number to obtain) implies no nuisance fuse operations without a circuit breaker operation. It is difficult for an outage data management system to properly determine the number of temporary faults.

Faults frequently occur near the peak of the voltage waveform as shown in Figure 4.5. About 60% of the faults in the EPRI fault study occurred when the voltage was within 5% of the peak prefault voltage (where the angle was

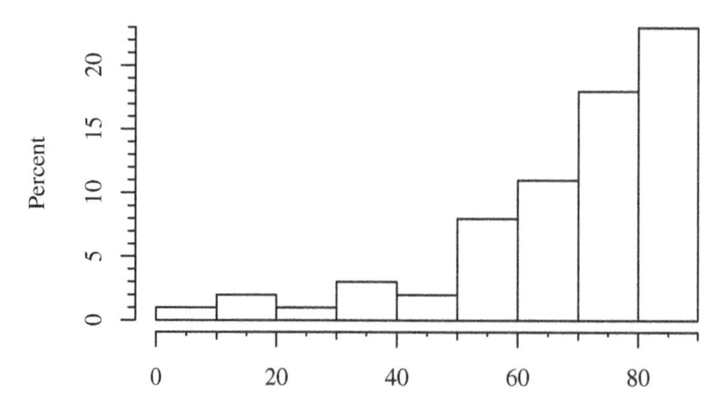

FIGURE 4.5
Point of fault on the voltage waveform. (Data from [Burke and Lawrence 1984; EPRI 1209-1, 1983].)

70 to 90°). This is reasonable. Any insulation failure, whether it be a squirrel breaching a bushing or a failure in a cable, more likely strikes with the voltage at or near its peak. Some faults defy this pattern. Lightning faults happen at any point on the voltage waveform because the fault occurs when the lightning strikes (although lightning can cause a flashover but not a fault if the voltage is very close to a zero-crossing of the power-frequency voltage). Two-phase and three-phase faults create more instances in which the voltage is not near its peak.

4.2 Fault Calculations

The magnitude of fault current is limited only by the system impedance and any fault impedance. The system impedance includes the impedances of wires, cables, and transformers back to the source. For faults involving ground, the impedance includes paths through the earth and through the neutral wire. The impedance of the fault depends on the type of fault.

Most distribution primary circuits are radial, with only one source and one path for fault currents. Figure 4.6 shows equations for calculating fault currents for common distribution faults.

The equations in Figure 4.6 assume that the positive-sequence impedance is equal to the negative-sequence impedance. As an example, the impedance term due to the sequence components for a line-to-line fault is $(Z_1 + Z_2)$, which simplifies to $2Z_1$ when the impedances are assumed to be equal. This is accurate for virtually all distribution circuits. With a large generator nearby, the equivalent circuit may have different positive- and negative-

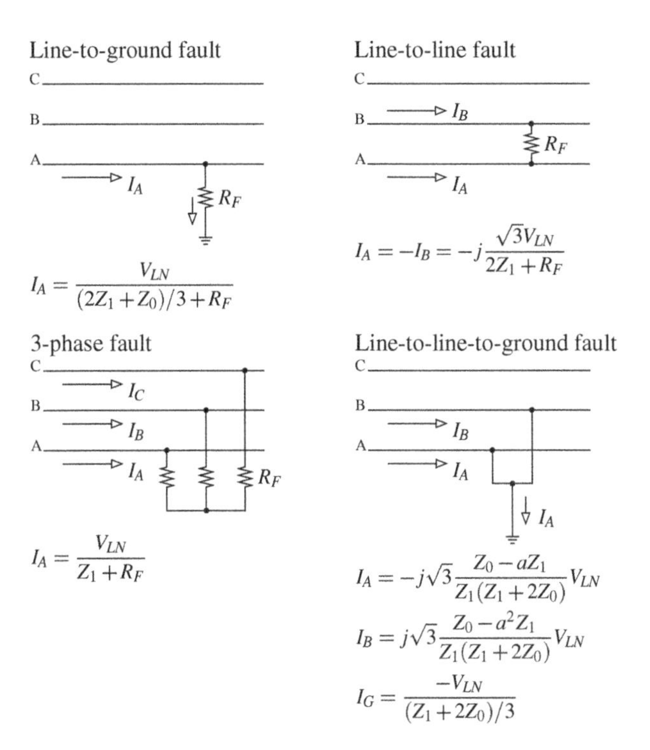

FIGURE 4.6
Fault-current calculations.

sequence impedances (but that case is usually done on the computer and not with hand calculations). The maximum currents occur with a bolted fault where R_F is zero. The maximum current for a line-to-line fault is 86.6% of the maximum three-phase fault current. In all cases, the load current is ignored. In most cases, load will not significantly change results.

The three-phase fault current is almost always the highest magnitude. On most circuits, the zero-sequence impedance is significantly higher than the positive-sequence impedance. One important location where the line-to-ground fault current may be higher is at the substation. There are two reasons for this:

1. A delta – wye transformer is a zero-sequence source. The positive-sequence impedance includes the impedance of the subtransmission and transmission system. The zero-sequence impedance does not. Figure 4.7 shows the sequence diagrams for the positive and zero sequences. The delta-wye connection forms a zero-sequence source while the positive-sequence impedance includes the subtransmission equivalent impedance.
2. If the substation transformer has three-legged core-form construction, the zero-sequence impedance is lower than its positive-

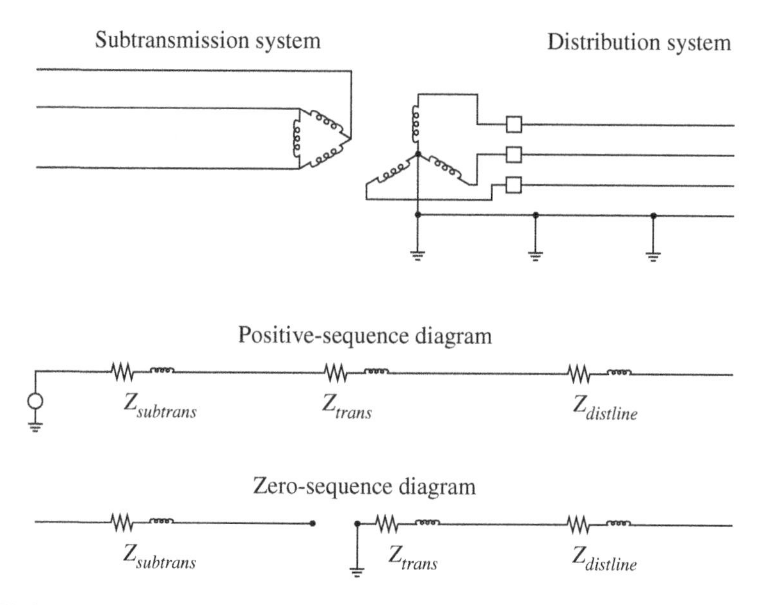

FIGURE 4.7
Positive and zero-sequence diagrams for a delta – wye substation transformer.

sequence impedance. Typically, the zero-sequence impedance is 85% of the positive-sequence impedance, which increases ground-fault currents by 5.2%.

In cases where the zero-sequence impedance is less than the positive-sequence impedance, the line-to-ground fault gives the highest phase current. The double line-to-ground fault produces the highest-magnitude ground current.

In order to reduce fault currents for line-to-ground faults, a neutral reactor on the station transformer is sometimes used. Figure 4.8 shows the equations for faults involving ground for circuits with a neutral reactor (the line-to-line fault and the three-phase fault are not affected). A common value for a neutral reactor is 1 Ω for 15-kV class distribution circuits.

The impedance seen by line-to-ground faults is a function of both the positive and the zero sequence impedances. This important loop impedance is $Z_S = (2Z_1 + Z_0)/3$. The sequence impedances, Z_1 and Z_0, used in the fault calculations include the sum of the impedances with both resistance and reactance along the fault current path. Some of the common branch impedances are given below including some rule-of-thumb values that are useful for hand calculations:

- Overhead lines:
  - $|Z_1| = 0.5\ \Omega/\text{mi}\ (0.3\ \Omega/\text{km})$
  - $|Z_S| = 1\ \Omega/\text{mi}\ (0.6\ \Omega/\text{km})$

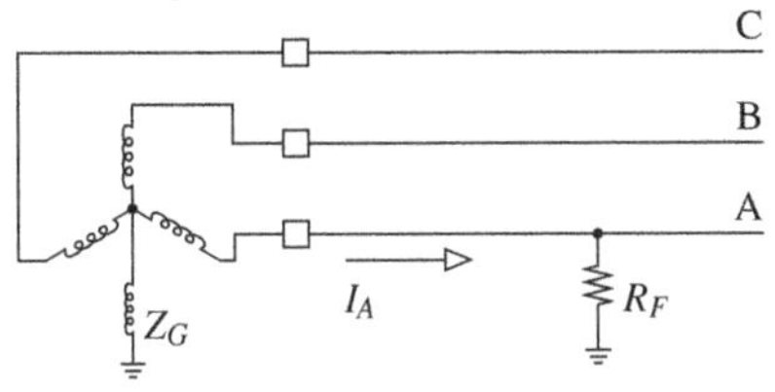

$$I_A = \frac{V_{LN}}{(2Z_1 + Z_0)/3 + R_F + Z_G}$$

Line-to-line-to-ground fault

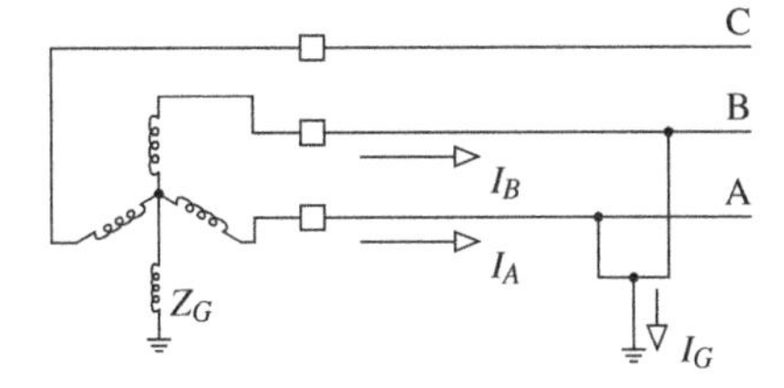

$$I_A = -j\sqrt{3}\frac{Z_0 + 3Z_G - aZ_1}{Z_1(Z_1 + 2Z_0 + 6Z_G)}V_{LN}$$

$$I_B = j\sqrt{3}\frac{Z_0 + 3Z_G - a^2Z_1}{Z_1(Z_1 + 2Z_0 + 6Z_G)}V_{LN}$$

$$I_G = \frac{-V_{LN}}{(Z_1 + 2Z_0 + 6Z_G)/3}$$

FIGURE 4.8
Fault-current calculations with a neutral reactor on the substation transformer.

- Underground cables:
  - $|Z_1|$ = 0.6 Ω/mi (0.35 Ω/km)
  - $|Z_S|$ = 0.5 Ω/mi (0.3 Ω/km)
- Substation transformer:
  - $|Z_1| = |Z_S| = 1\ \Omega$
  - A typical 15-kV substation transformer impedance is 1 Ω, which corresponds to a bus fault current of 4.2 kA for a 12.47-kV circuit.
- Subtransmission equivalent: often can be ignored

See Smith (1980) for an excellent paper on fault calculations for additional information. Include impedances for step-down transformer banks, series reactors, and voltage regulators. Use the rule-of-thumb numbers above for back-of-the-envelope calculations and as checks for computer modeling.

The simplified equation for a transformer impedance is

$$Z_1 = Z_0 = j\frac{kV^2}{MVA}Z_{\%}$$

where

kV = line-to-line voltage
MVA = transformer base rating — open air (OA) rating
$Z_{\%}$ = transformer impedance, per unit

We ignore the resistive component since the *X/R* ratio of station transformers is generally greater than 10 and often in the range of 20 to 30. The transmission/subtransmission equivalent is usually small, and we often ignore it (especially for calculating maximum fault currents). Include the transmission system impedance for weak subtransmission systems such as 34.5-, 46- or 69-kV circuits, or for very large substations. Find the transmission equivalent from the per unit impedances (r_1, x_1, r_0, and x_0) on a given MVA base referred to the distribution voltage (Smith, 1980) as

$$Z_1 = (r_1 + jx_1)\frac{kV_s^2}{MVA_b}\left(\frac{kV_{pb}}{kV_p}\right)^2$$

$$Z_0 = (r_0 + jx_0)\frac{kV_s^2}{MVA_b}\left(\frac{kV_{pb}}{kV_p}\right)^2$$

where

MVA_b = base MVA at which the r and x impedances are given
kV_s = line-to-line voltage in kV on the secondary side of the station transformer
kV_p = line-to-line voltage in kV on the primary
kV_{pb} = base line-to-line voltage on the primary used to calculate MVA_b (often equal to kV_p)

If the transmission impedances are available as a fault MVA with a power factor, find the transmission equivalent (Smith, 1980) with

$$Z_1 = \frac{kV_s^2}{MVA}\left(pf + j\sqrt{1 - pf^2}\right)\left(\frac{kV_{pb}}{kV_p}\right)^2$$

$$Z_0 = \frac{\sqrt{3}kV_s^2}{kI_g \cdot kV_{pb}}\left(pf_g + j\sqrt{1 - pf_g^2}\right)\left(\frac{kV_{pb}}{kV_p}\right)^2 - 2Z_1$$

where

MVA = 3-phase short-circuit MVA at the primary terminals of the station transformer (see Table 4.2 for typical maximum values)
kI_g = available ground fault current in kA at the primary terminals of the station transformer
pf = power factor in per unit for the available three-phase fault current
pf_g = power factor in per unit for the available single-phase fault current

While almost all distribution circuits are radial, there may be other fault current sources. We ignore these other sources most of the time, but occasionally, we consider motors and generators in fault calculations. Synchronous motors and generators contribute large currents relative to their size. On a typical 15-kV class distribution circuit, one or two megawatts worth

TABLE 4.2

Typical Maximum Transmission/Subtransmission Fault Levels

Transmission Voltage, kV	Maximum Symmetrical Fault, MVA
69	3,000
115	5,000
138	6,000
230	10,000

of connected synchronous units are needed to significantly affect fault currents. On weaker circuits, smaller units can impact fault currents. Induction motors and generators also feed faults. Inverter-based distributed generation can contribute fault current, but generally much less than synchronous or induction units. Of course, on feeders that have network load, current through network transformers backfeeds faults until the network protectors operate.

4.2.1 Transformer Connections

The fault current on each side of a three-phase transformer connection can differ in magnitude and phasing. In the common case of a delta – grounded-wye connection, the current on the source side of the transformer differs from the currents on the fault side for line-to-ground or line-to-line faults (see Figure 4.9). For a line-to-ground fault on the primary side of the transformer, the current appears on two phases on the primary with a per unit current of 0.577 (which is $1/\sqrt{3}$).

These differences are often needed when coordinating a primary-side protective device and a secondary-side device. In distribution substations, this is commonly a fuse on the primary side and a relay controlling a circuit breaker on the secondary side. The line-to-line fault must be considered — this gives more per-unit current on one phase in the primary, 1.15 per unit

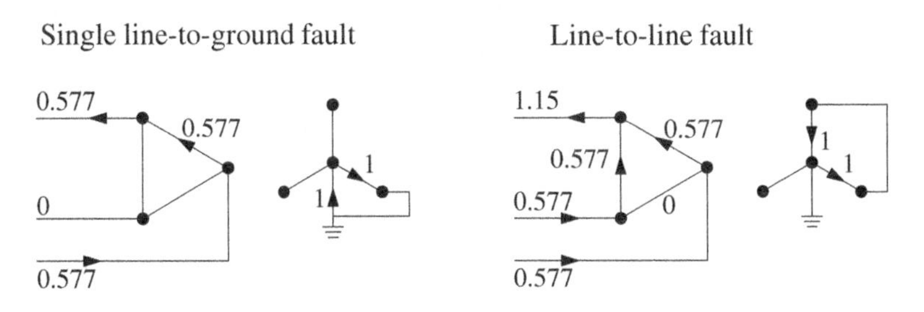

FIGURE 4.9
Per-unit fault-currents on both sides of a delta – grounded-wye transformer.

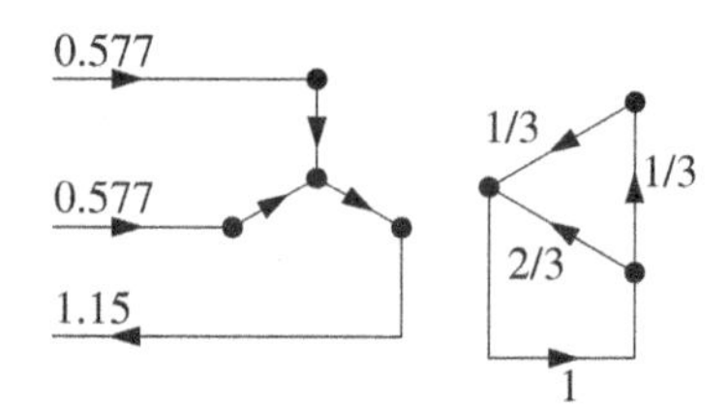

FIGURE 4.10
Per-unit fault-currents on both sides of a wye – delta transformer.

($2/\sqrt{3}$) in one of the phases (see Figure 4.9). To make sure a primary fuse coordinates with a secondary device, shift the minimum-melting time-current curve of the primary-side fuse to the left by a factor of 0.866 = $\sqrt{3}/2$ (after also adjusting for the transformer turns ratio). The current differences also mean that the transformer is not protected as well for single-phase faults; a primary-side fuse takes longer to clear the single-phase fault since it sees less current than for a three-phase or line-to-line fault.

Fault currents are only different for unbalanced secondary currents. For a three-phase secondary fault, the per-unit currents on the primary equal those on the secondary (with the actual currents related by the turns ratio of the transformer). A wye – wye transformer does not disturb the current relationships; the per-unit currents on both sides of the transformer are equal.

In a floating wye – delta, similar current relationships exist; a line-to-line secondary fault shows up on the primary side on all three phases, one of which is 1.15 per unit (see Figure 4.10). For a floating wye – delta transformer with a larger center-tapped lighting leg and two power legs, fault current calculations are difficult. Faults can occur from phase to phase and from phase to the secondary neutral, and the lighting transformer will have a different impedance than the power leg transformers. For an approach to modeling this, see Kersting and Phillips (1996).

4.2.2 Fault Profiles

Fault profiles show fault current with distance along a circuit. Determining where thermal or mechanical short-circuit limits on equipment may be exceeded, helping select or check interrupting capabilities of protective equipment, and coordinating protective devices are important uses of fault profiles. Figure 4.11 and Figure 4.12 show typical fault current profiles of distribution circuits.

Some general trends that the fault profiles show are:

- *Distance* — The fault current drops off as the inverse of distance ($1/d$).
- *Ground faults* — On overhead circuits, the ground fault current falls off faster (and the ground fault current is generally lower)

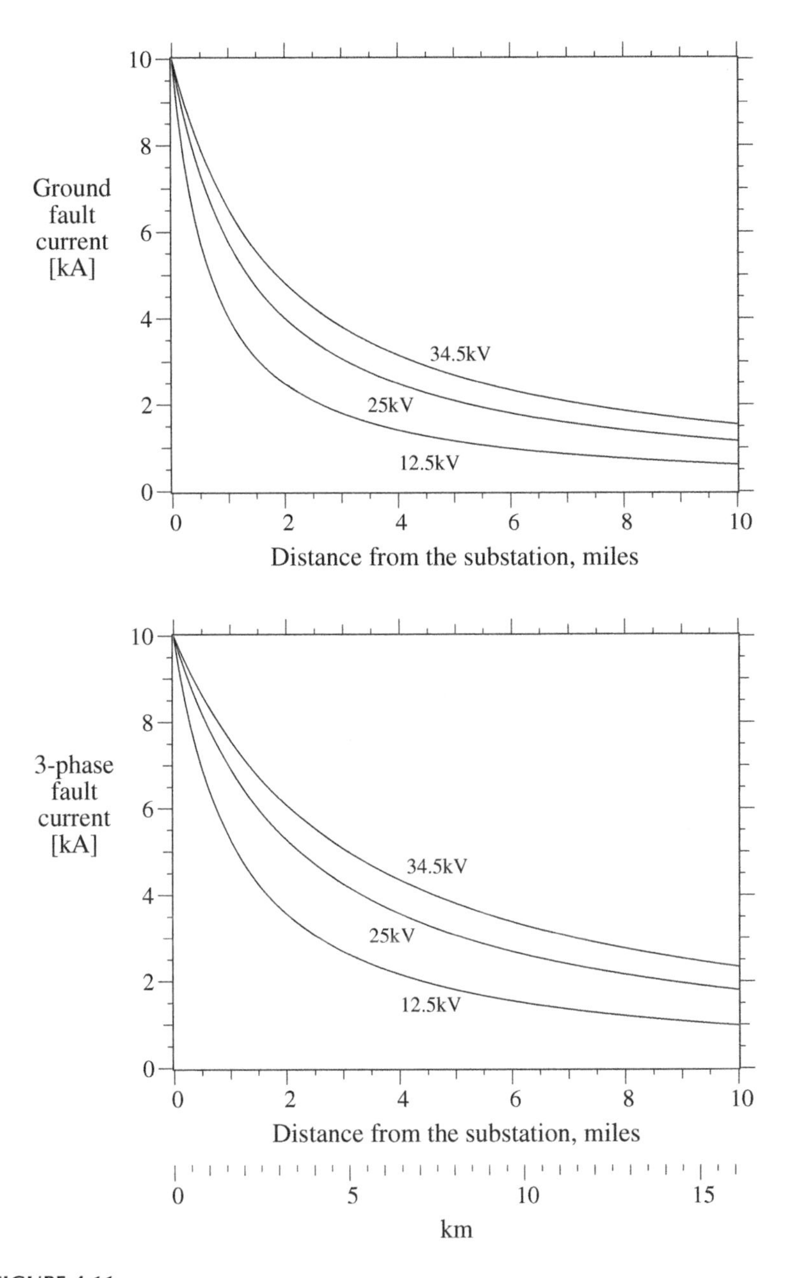

FIGURE 4.11
Fault-current profiles for line-to-ground faults and for three-phase faults for an overhead circuit. Phase characteristics: 500 kcmil, all-aluminum, GMD = 4.69 ft (1.43 m). Neutral characteristics: 3/0 all-aluminum, 4-ft (1.22-m) line-neutral spacing. $Z_1 = 0.207 + j0.628\ \Omega$/mile ($0.1286 + j0.3901\ \Omega$/km), $Z_0 = 0.720 + j1.849\ \Omega$/mile ($0.4475 + j1.1489\ \Omega$/km), $Z_S = 0.378 + j1.035\ \Omega$/mile ($0.2350 + j0.6430\ \Omega$/km).

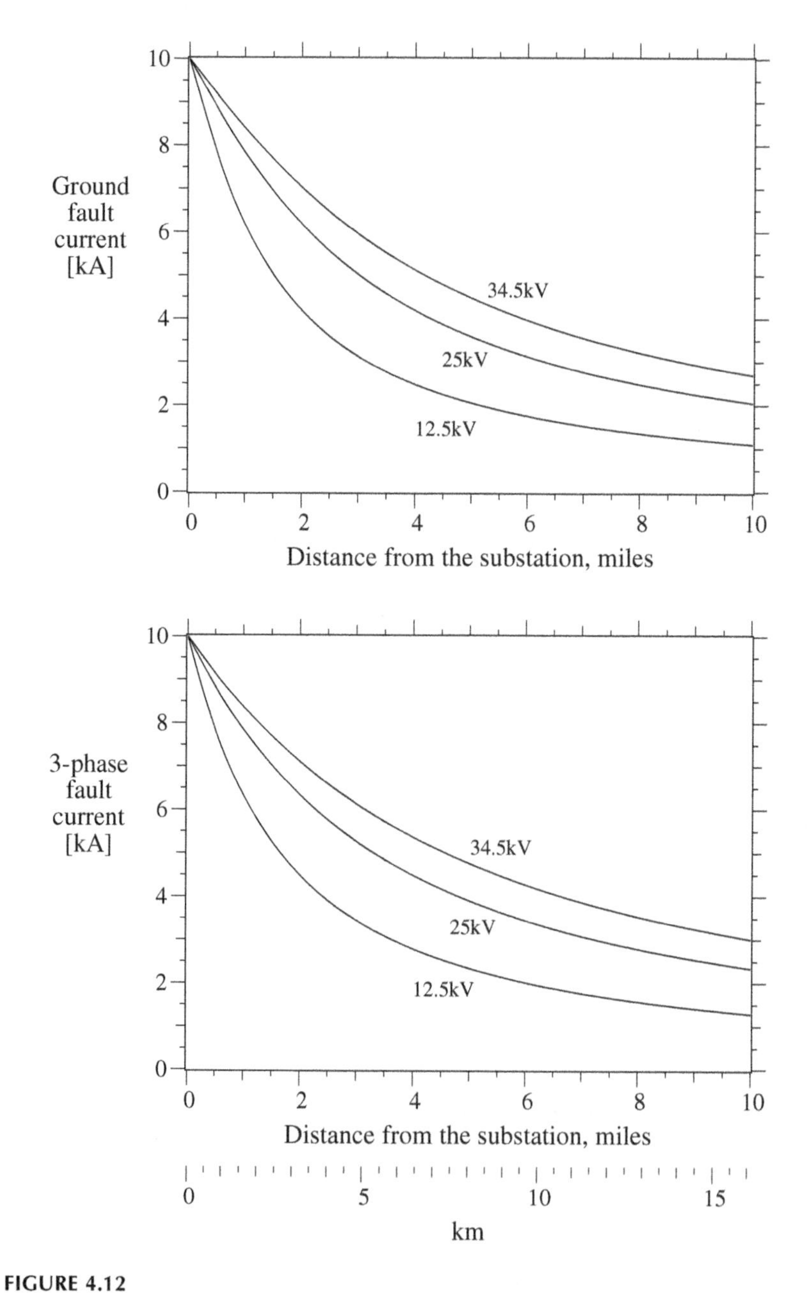

FIGURE 4.12
Fault-current profiles for line-to-ground faults and for three-phase faults for an underground cable circuit. 500-kcmil aluminum conductor, 220-mil XLPE insulation, 1/3 neutrals, flat spacing, 4.5 in. between cables. $Z_1 = 0.3543 + j0.3596\ \Omega/\text{mile}$ ($0.2201 + j0.2234\ \Omega/\text{km}$), $Z_0 = 0.8728 + j0.2344\ \Omega/\text{mile}$ ($0.5423 + j0.1456\ \Omega/\text{km}$), $Z_S = 0.5271 + j0.3178\ \Omega/\text{mile}$ ($0.3275 + j0.1975\ \Omega/\text{km}$).

than the three-phase fault current. The zero-sequence reactance is generally over three times the positive-sequence reactance, and the zero-sequence resistance is also higher than the positive-sequence resistance.

- *System voltage* — On higher-voltage distribution systems, the fault current drops off more slowly. The actual line impedance does not change with voltage ($Z_S \approx 1\ \Omega/\text{mi}$), and since $I = V_{LN}/Z$, it takes more impedance (more circuit length) to reduce the fault current.
- *Cables* — Underground cables have much lower reactance than overhead circuits, so the fault current does not fall off as fast on underground circuits. Also, note that *X/R* ratios are lower on cables.
- *Profiles* — The three-phase and ground-fault profiles of underground cables are similar. The zero-sequence reactance can actually be smaller than the positive-sequence reactance (but the zero-sequence resistance is larger than the positive-sequence resistance).

4.2.3 Effect of *X/R* Ratio

In a reactive circuit (high *X/R* ratio), it is naturally more difficult for a protective device such as a circuit breaker to clear a fault. Protective devices clear a fault at a current zero. Within the interruptor, dielectric strength builds up to prevent the arc from reigniting after the current zero. In a resistive circuit (low *X/R* ratio), the voltage and current are in phase, so after a current zero, a quarter cycle passes before the voltage across the protective device (called the *recovery voltage*) reaches its peak. In a reactive circuit, the fault current naturally lags the voltage by 90°; the voltage peaks at a current zero. Therefore, the recovery voltage across the protective device rises to its peak in much less than a quarter cycle (possibly in 1/20th of a cycle or less), and the fault arc is much more likely to reignite.

Another factor that makes it more difficult for protective devices to clear faults is asymmetry. Circuits with inductance resist a change in current. A short circuit creates a significant change in current, possibly creating an offset. If the fault occurs when the current would naturally be at its negative peak, the current starts at that point on the waveshape but is offset by 1.0 per unit. The dc offset decays, depending on the *X/R* ratio. The offset is described by the following equation:

$$i(t) = \underbrace{\sqrt{2}I_{rms}\sin(2\pi f t + \beta - \theta)}_{\text{ac component}} - \underbrace{\sqrt{2}I_{rms}\sin(\beta - \theta)e^{-\frac{2\pi f t}{X/R}}}_{\text{decaying dc component}}$$

where

$i(t)$ = instantaneous value of current at time t

I_{rms} = root-mean square (rms) value of the ac component of current,

$$I_{rms} = V / \sqrt{R^2 + X^2}$$

β = the closing angle which defines the point on the waveform at which the fault is initiated

θ = system impedance angle = $\tan^{-1}\frac{X}{R}$

f = system frequency, Hz

t = time, sec

Asymmetry is higher with higher X/R ratios. The worst case offset with $X/R = \infty$ is 2 per unit. Figure 4.13 shows an example of an offset fault current.

If a phase faults at the natural zero crossing ($\beta = \theta$), no offset occurs. The highest magnitude of the dc component occurs when the fault happens 90° from the natural zero crossing of the circuit (when $\beta = \theta \pm \pi/2$). The highest

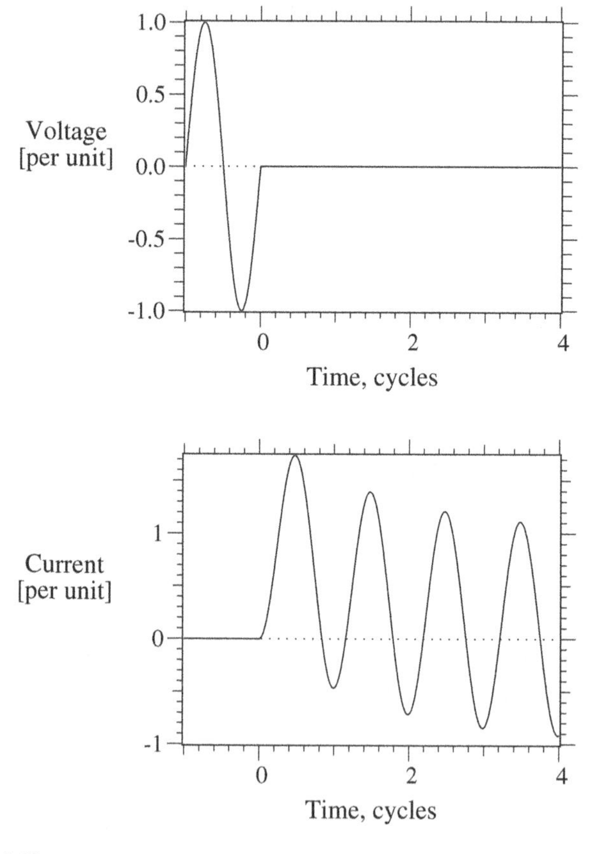

FIGURE 4.13
Example of an asymmetric fault with $X/R = 10$ which initiated when the closing angle $\beta = 0$, which is when the voltage crosses zero.

dc offset does not align with the highest peak asymmetric current (which is the sum of the ac and decaying dc component). The peak current occurs when the closing angle $\beta = 0$ for all *X/R* ratios ($\beta = 0$ when the fault occurs at a voltage zero crossing). The ratio of the peak current I_p to the rms current I_{rms} can be approximated by

$$\frac{I_p}{I_{rms}} = \sqrt{2}\left[1 + e^{-(\theta+\pi/2)\frac{R}{X}} \sin\theta\right] = \sqrt{2}\left[1 + e^{-\pi\frac{R}{X}}\right] \quad \text{for } \theta = \pi/2$$

This is the most industry-accepted approximation that is used, but it gives an approximation that is slightly low. A more accurate approximation can be found (St. Pierre, 2001) with

$$\frac{I_p}{I_{rms}} = \sqrt{2}\left[1 + e^{-2\pi\frac{R}{X}\tau}\right]$$

where τ is a fictitious time found with

$$\tau = 0.49 - 0.1e^{-\frac{1}{3}\frac{X}{R}}$$

In addition to causing a higher peak magnitude, asymmetry also causes a longer first half cycle (important for fuse operating time) and much higher first half cycle $\int I^2\,dt$. The occurrence of asymmetry is reduced by the fact that most faults occur when the voltage is near its peak (Figure 4.5). In a circuit with a high *X/R* ratio, when the voltage is at its peak, the fault current is naturally near zero. Therefore, for most faults, the asymmetry is small, especially for line-to-ground faults. For two- or three-phase faults where each phase is faulted simultaneously (as can happen with lightning), asymmetry is much more likely.

Asymmetry is important to consider for application of cutouts, circuit breakers, and other equipment with fault current ratings. Equipment is generally tested at a given *X/R* ratio. If the equipment is applied at a location where the *X/R* ratio is higher, then the equipment may have less capability than the rating indicates. Equipment often has a momentary duty rating which is the short-time (first-cycle) withstand capability. This is strongly influenced by asymmetry.

Other impacts of asymmetry include:

- Asymmetry can saturate current transformers (CTs). On distribution circuits, overcurrent relays should still operate although they could be more susceptible to miscoordination.
- Fuses respond to $\int I^2 dt$, so asymmetrical current melts the link significantly faster.

- Asymmetry can foul up fault-location algorithms in digital relays and fault recorders.

4.2.4 Secondary Faults

Secondary faults vary depending on the transformer connection and the type of fault on the secondary. For a standard single-phase 120/240-V secondary for residential service, two faults are of interest: a fault from a phase to the neutral and a fault from one of the hot legs to the other across the full 240 V. The impedance to the fault includes the transformer plus the secondary impedance. The secondary current for a bolted fault across the 240-V legs (between the two hot legs) is

$$I_{240} = \frac{240}{\sqrt{\left(R_T + \frac{R_S L}{1000}\right)^2 + \left(X_T + \frac{X_S L}{1000}\right)^2}}$$

where

I_{240} = Secondary current, symmetrical A rms for a 240-V fault (phase-to-phase)
R_T = Transformer full-winding resistance, Ω at 240 V (from terminals X1 to X3)
X_T = Transformer full-winding reactance, Ω at 240 V (from terminals X1 to X3)
R_S = Secondary conductor resistance to a 240-V fault, Ω/1000 ft
X_S = Secondary conductor resistance to a 240-V fault, Ω/1000 ft
L = Distance to the fault, ft

and

$$R_T = 0.0576 \frac{W_{CU}}{S_{kVA}^2}$$

$$Z_T = 0.576 \frac{Z_{\%}}{S_{kVA}}$$

$$X_T = \sqrt{Z_T^2 - R_T^2}$$

where

S_{kVA} = transformer rating, kVA
$W_{CU} = W_{TOT} - W_{NL}$ = load loss at rated load, W
W_{TOT} = total losses at rated load, W
W_{NL} = no-load losses, W
$Z_{\%}$ = nameplate impedance magnitude, %

For a short circuit from one of the hot legs to the neutral, both the transformer and the secondary have different impedances. For the transformer, the half-winding impedance must be used; for the secondary, the loop impedance through the phase and the neutral should be used.

$$I_{120} = \frac{120}{\sqrt{\left(R_{T1} + \frac{R_{S1}L}{1000}\right)^2 + \left(X_{T1} + \frac{X_{S1}L}{1000}\right)^2}}$$

where

I_{120} = Secondary current in symmetrical A rms for a 120-V fault (phase-to-neutral)

R_{T1} = Transformer half-winding resistance, Ω at 120 V (from terminals X1 to X3)

X_{T1} = Transformer half -winding reactance, Ω at 120 V (from terminals X1 to X3)

R_{S1} = Secondary conductor resistance to a 120-V fault, Ω/1000 ft

X_{S1} = Secondary conductor resistance to a 120-V fault, Ω/1000 ft

L = Distance to the fault, ft

In absence of better information, use the following impedances for transformers with an interlaced secondary winding:

$$R_{T1} = 0.375R_T \quad \text{and} \quad X_{T1} = 0.3X_T$$

And use the following impedances for transformers with noninterlaced secondary windings:

$$R_{T1} = 0.4375R_T \quad \text{and} \quad X_{T1} = 0.625X_T$$

Figure 4.14 shows fault profiles for secondary faults on various size transformers. The secondary is triplex with 3/0 aluminum conductors and a reduced neutral. It has impedances of

$$R_S = 0.211\ \Omega/1000\ \text{ft} \qquad X_S = 0.0589\ \Omega/1000\ \text{ft}$$

$$R_{S1} = 0.273\ \Omega/1000\ \text{ft} \qquad X_{S1} = 0.0604\ \Omega/1000\ \text{ft}$$

The secondary has significant impedances; fault currents drop quickly from the transformers. Close to the transformer, line-to-neutral faults are higher magnitude. At large distances from the transformer, the secondary impedances dominate the fault currents. Faults across 240 V are normally higher magnitude than line-to-neutral faults.

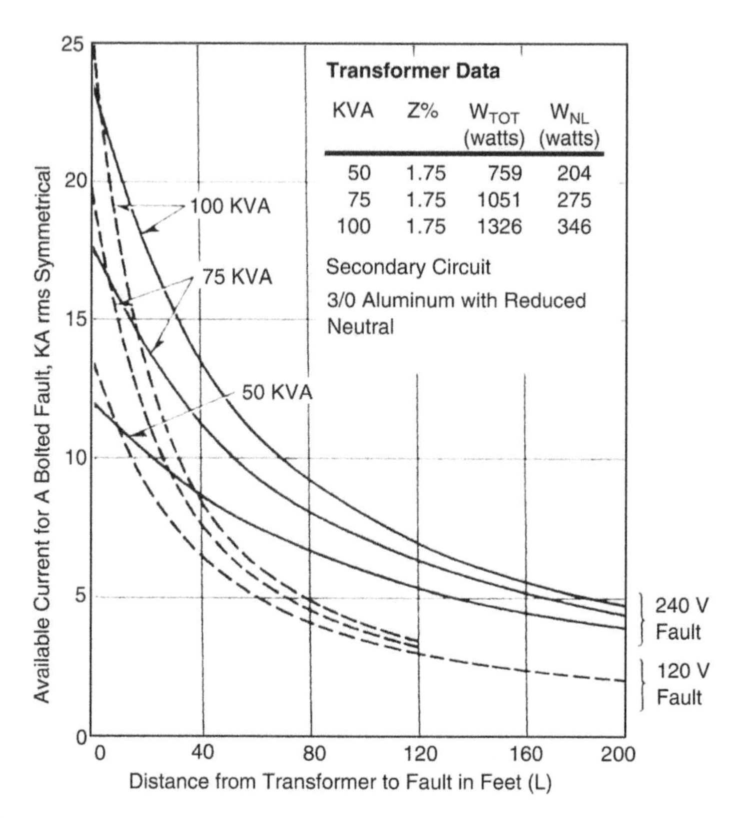

FIGURE 4.14
Fault profiles for 120/240-V secondary faults ($R_{T1} = 0.375R_T$, and $X_{T1} = 0.5X_T$). (From ABB Inc., *Distribution Transformer Guide*, 1995. With permission.)

Normally in secondary calculations, we can ignore the impedance offered by the distribution primary. The primary-system impedance is usually small relative to the transformer impedance, and neglecting it is conservative for most uses. On weak distribution systems or with large, low-impedance distribution transformers, the distribution system impedance plays a greater role.

4.2.5 Primary-to-Secondary Faults

Faults from the distribution primary to the secondary can subject end-use equipment to significant overvoltages. Figure 4.15 shows a circuit diagram of a fault from the primary to a 120/240-V secondary. This type of fault can occur several ways: a high-to-low fault within the transformer, a broken primary wire falling into the secondary, or a broken primary jumper. As we will discuss, the transformer helps limit the overvoltage. Having the primary fall on the secondary does not automatically mean primary-scale voltages in customers' homes and facilities.

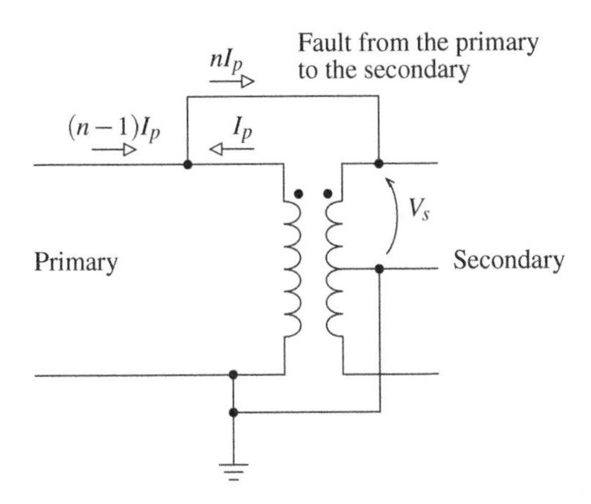

FIGURE 4.15
Fault from the primary to a 120/240-V secondary circuit.

The per-unit secondary voltage for a fault from the primary to the secondary (PTI, 1999) is

$$V_s = \frac{n}{1+(n-1)^2 \dfrac{S_{kVA}}{10\ V_{kV}\ I_{kA}\ Z_{\%}}}$$

where
V_s = secondary voltage, per unit at 120 V
n = transformer turns ratio from the primary voltage to the half-voltage secondary rating (normally 120 V)
I_{kA} = available primary fault current for a single line-to-ground fault, kA
S_{kVA} = transformer rating, kVA
$Z_{\%}$ = half winding impedance of the transformer, %
V_{kV} = primary line-to-ground winding rated voltage, kV

Figure 4.16 shows the per-unit overvoltage for various transformer sizes. Surprisingly, the primary voltage does not impact the overvoltage significantly. The overvoltage equation in per unit reduces (PTI, 1999) to approximately

$$V_s \approx \frac{1.2 Z_{\%} I_{kA}}{S_{kVA}}$$

The overvoltage increases with higher available fault current, higher impedance transformers, and smaller transformers. For all but the smallest transformers with the highest impedance, the overvoltage is not too hazardous. But, if a fuse operates to separate the transformer from the circuit but

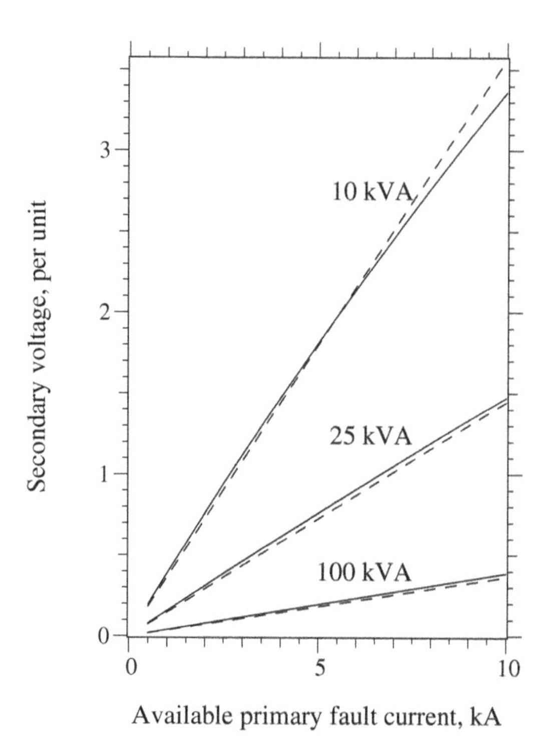

FIGURE 4.16
Secondary voltage during a fault from the primary to a 120/240-V secondary circuit. The solid lines are for a 4.8-kV circuit, and the dashed lines are for a 34.5-kV circuit. The results assume that $Z_{\%} = 3\%$.

leaves the primary-to-secondary fault, the fault imposes full primary voltage on the secondary (at least until the first failure on the secondary system). Such a condition can occur when the fault starts on the primary side above the transformer fuse (see Figure 4.17). If the transformer fuse blows before the upstream line fuse, the secondary voltage rises to the primary voltage. If the fault is below the transformer fuse, it does not matter which fuse blows first; either clears the fault.

The example in Figure 4.15 shows a fault to the secondary leg that is in phase with the primary (off of the *X*1 bushing of the transformer). A fault to the other secondary leg (off of *X*3) has very similar effects; the voltages and currents are almost the same, so the equations and graphs in this section also apply.

Although the transformer helps hold down the overvoltage, the primary-to-secondary fault may initiate a sizeable switching transient that could impact end-use equipment.

With most line fuses and transformer fuses used, the line fuse will clear before the transformer fuse and before the transformer suffers damage (good news on both counts). Even though the upstream fuse is larger, it sees $(n - 1)$

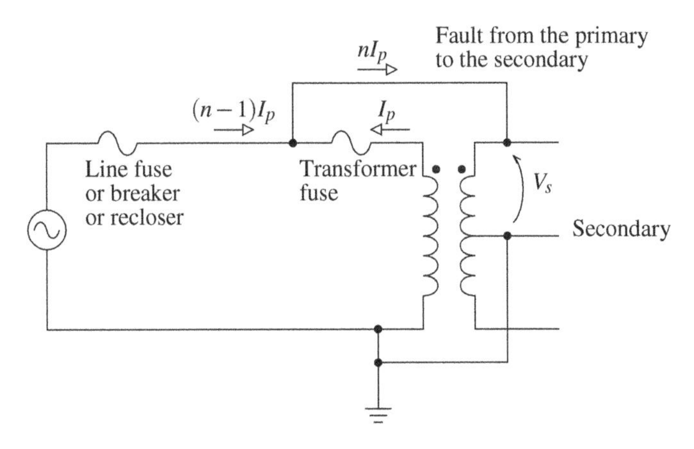

FIGURE 4.17
Fault from the primary to a 120/240-V secondary circuit.

times the fault current. With the primary fault above the transformer fuse, the transformer fuse is more likely to operate before the line fuse with

- *Small transformer fuses* — Another reason not to fuse transformers too tightly; smaller, fast transformer fuses are more likely to clear before an upstream device.
- *Upstream breaker or recloser* — If the upstream device is a circuit breaker or recloser instead of a fuse, the tripping time is much longer, especially on a time-delayed trip (but even a fast trip is relatively long). If a circuit breaker is upstream of the transformer, the transformer fuse is likely to blow before the circuit breaker for locations with high fault currents and with small transformer fuses.

A more detailed analysis of the coordination of the two devices requires using the time-current characteristics of each of the protective devices along with the currents. The current into the primary winding, I_p in kA is

$$I_p = \frac{I_{kA}}{(n-1) + \frac{10 I_{kA}}{(n-1)} \frac{V_{kV}}{S_{kVA}} Z_{\%}} \approx \frac{I_{kA}}{n}$$

Again, the upstream device sees $(n-1)I_p$, which is almost the full available current for a single line-to-ground fault, I_{kA}.

A transformer with a secondary circuit breaker (as in a completely self-protected transformer, a CSP) has another possible mode where the transformer separates. If the secondary circuit breaker opens before the upstream primary device, the high-to-low fault raises the secondary voltage to the primary voltage. The secondary circuit breaker may not be able to clear the

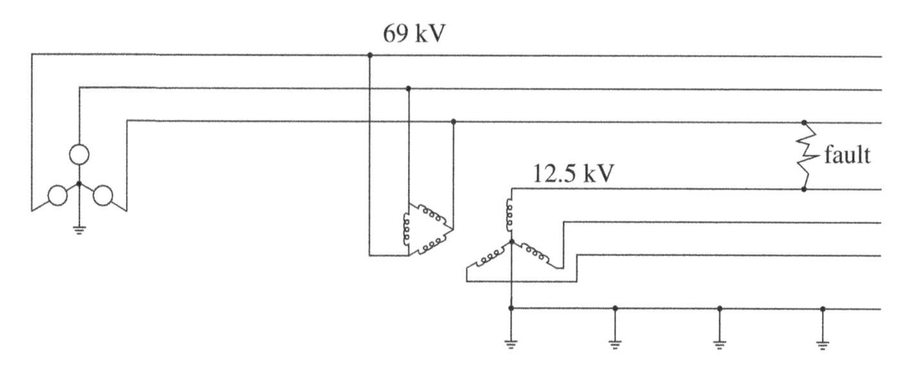

FIGURE 4.18
Example of a fault from a transmission conductor to a distribution conductor.

fault because the arc recovery voltage is much higher than the rating for the secondary circuit breaker; this is good news in that it helps protect end-use equipment from extreme overvoltages, but the secondary circuit breaker may fail trying to clear the fault. If the upstream device is a fuse, the fuse will probably clear before the secondary circuit breaker opens, but if the upstream device is a circuit breaker, the secondary circuit breaker will probably try to open first.

4.2.6 Underbuilt Fault to a Transmission Circuit

Faults from transmission circuits to distribution circuits are another hazard that can subject distribution equipment and customer equipment to extremely high voltages. Consider the example in Figure 4.18 of a fault from a subtransmission circuit to a distribution circuit.

As is the case for primary-to-secondary faults discussed in the previous section, overvoltages are not extremely high as long as the distribution circuit stays connected. But if a distribution interrupter opens the circuit, the voltage on the faulted distribution conductor jumps to the full transmission-line voltage. With voltage at several times normal, something will fail quickly. Such a severe overvoltage is also likely to damage end-use equipment. The distribution interrupter, either a circuit breaker or recloser, may not be able to clear the fault (the recovery voltage is many times normal); it may fail trying.

Faults further from the distribution substation cause higher voltages, with the highest voltage at the fault location. Current flowing back towards the circuit causes a voltage rise along the circuit.

While one can use a computer model for an exact analysis (but it is not possible with most standard distribution short-circuit programs), a simplified single-phase analysis (assuming a wye – wye transformer) helps frame the problem. The fault current is approximately

$$I = \frac{V_S}{\frac{(n-1)}{n} Z_A + \frac{n}{(n-1)} Z_B} \approx \frac{V_S}{Z_A + Z_B}$$

where

n = ratio of the transmission to distribution voltage ($n = 69/12.5 = 5.5$ in the example)

V_S = rms line-to-ground transmission source voltage (40 kV in the example)

Z_A = loop impedance from the transmission source to the high side of the distribution station

Z_B = loop impedance from the high-side at the distribution station out to the fault and back to the distribution low-side of the distribution substation

And, the 69-kV impedance often dominates, so the fault current is really determined by Z_A. For the distribution and transmission line impedances, Z_A and Z_B, you can use 1 Ω/mi for quick approximations. The worst case is with a small Z_A, a stiff subtransmission system.

The voltage at the fault is

$$V = I\frac{Z_B}{2} + V_d$$

where

V_d = line-to-ground voltage on the distribution circuit at the substation (as a worst case, assume that it is the nominal voltage; it will usually be less because of the sag that pulls down the voltages).

Figure 4.19 shows results from a series of computer simulations on a 12.5-kV circuit for various fault locations and subtransmission source stiffnesses. Results only modestly differ for other configurations: a 69-kV source in the opposite direction, a looped transmission source, a different substation transformer configuration, or different phases faulted. The worst cases are for stiff transmission systems.

If a distribution interrupter opens to leave transmission voltage on the distribution circuit, distribution transformers would saturate and metal-oxide arresters would move into heavy conduction. Transformer saturation distorts the voltage but may not appreciably reduce the peak voltage. Arresters can reduce the peak voltage, but they could still allow quite high voltages to customers. Arresters with an 8.4-kV maximum continuous operating voltage start conducting for power-frequency voltages at about 11 to 12 kV (1.5 to 1.6 times the nominal system line-to-ground voltage). At higher voltages, the arresters will draw more current. Depending on the number of arresters, a stiff transmission source can still push the voltage to between 3 and 4 per

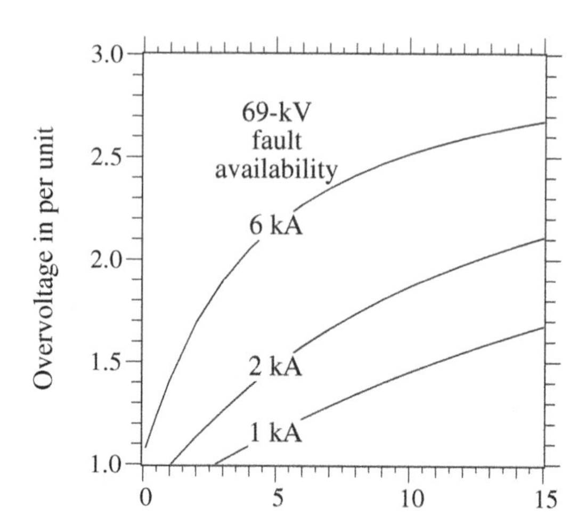

FIGURE 4.19
Results of simulations of a fault from a 69-kV circuit to a 12.5-kV circuit (before the distribution substation breaker trips).

unit, which is 20 to 30 kV (until an arrester or something else fails). In fact, the best protection happens when the arrester fails as fast as possible; the arrester becomes a sacrificial protector. Normal-duty arresters fail faster than heavy-duty arresters, which limits the duration of overvoltages. Goedde et al. (2002) propose using gapped arresters; their lab tests found that gapped arresters clip the overvoltage to a lower magnitude and fail faster during overvoltages.

To reduce the hazard of underbuilt distribution lines, consider the following experimental options:

- *Arresters* — Use normal-duty arresters and possibly gapped arresters.
- *Fuses* — Try to avoid using fuses or reclosers where it leaves significant downstream exposure underbuilt. The fast operation of fuses and reclosers are more likely to clear the distribution circuit before the overbuilt circuit.
- *Directional relays* — In faults to a transmission circuit, the power flows from the fault into the distribution station transformer (the opposite direction of power flow for normal faults). Tripping the distribution circuit breaker only for faults with forward power flow leaves the circuit breaker in for subtransmission faults.
- *Disable the instantaneous* — Without the instantaneous trip on the distribution feeder, the circuit breaker will wait longer before tripping. The transmission circuit is more likely to trip first.
- *Coordinate devices* — Coordinate the transmission-line protection to clear before the distribution circuit operates over the range of fault

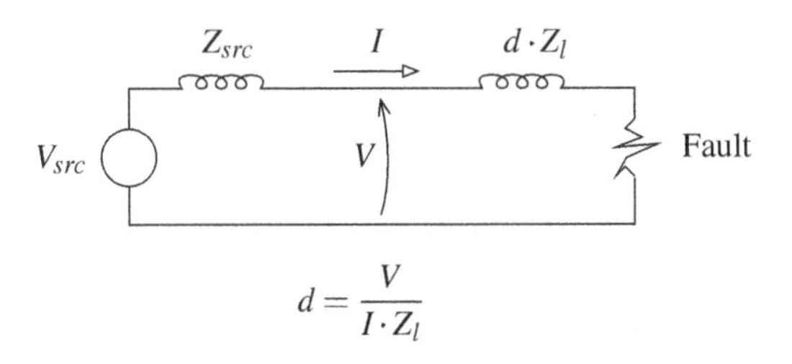

FIGURE 4.20
Fault location calculations.

currents that can occur. Include the effects of multiple reclose operations on the transmission circuit. Evaluate the substation recloser or circuit breaker; and also, consider feeder protective devices (normally reclosers). Try to speed up clearing times on the transmission circuit as much as possible (both ends for looped subtransmission circuits).

- *Ground switch (very experimental)* — Whenever the distribution circuit breaker opens, engage a grounding switch on the load side of the distribution circuit breaker. This grounds the fault, preventing overvoltages and sustaining the fault on the transmission circuit.
- *Structures* — As much as possible, design the common structure to minimize the chance of faults between circuits. Use wide spacings between the two circuits, and build the subtransmission circuit to high mechanical standards to reduce the chance of broken conductors or crossarms or braces. More extreme protection could be provided by stringing a grounded conductor and placing it between the subtransmission and the distribution circuits. If the ground is involved in the fault, it will prevent overvoltages. These options are obviously difficult to retrofit, but these issues should be considered when designing new circuits.

Thoroughly review such options before implementation. Normally, utilities treat underbuilt circuits the same as any other circuit. These are experimental approaches; I do not know of any implementation of these options for circuits with overbuilt transmission. Also, most of these options do not help as much for distribution lines fed from a different transmission source.

4.2.7 Fault Location Calculations

If we know the voltages and currents during a fault, we can use these to estimate the distance to the fault. The equation is very simple, just Ohm's Law (see Figure 4.20):

$$d = \frac{V}{I \cdot Z_l}$$

where

V = voltage during the fault, V

I = current during the fault, A

Z_l = line impedance, Ω/length unit

d = distance to the fault, length unit such as mi

With complex values entered for the voltages and impedances and currents, the distance estimate should come out as a complex number. The real component should be a realistic estimate of the distance to the fault; the imaginary component should be close to zero. If not, then something is wrong.

While the idea is simple, a useful implementation is more difficult. Different fault types are possible (phase-to-phase, phase-to-ground, etc.), and each type of fault sees a different impedance. Fault currents may have offsets. The fault may add impedance. There are uncertainties in the impedances, especially the ground return path. Conductor size changes also make location more difficult.

Many relays or power quality recorders or other instruments record fault waveforms. Some relays have fault-locating algorithms built in.

The Ohm's Law equation is actually overdetermined. We have more information than we really need. The distance is a real quantity, but the voltages, currents, and impedances are complex, so the real part of the result is the distance, and the reactive part is zero. Most fault-locating algorithms use this extra information, allow the fault resistance to vary, and find the distance that provides the optimal fit (Girgis et al., 1993; Santoso et al., 2000). The problem with this approach is that the fault resistance soaks up the error in other parts of the data. It does not necessarily mean a better distance estimation. Most fault arcs have a resistance that is very close to zero. In most cases, we're better off assuming zero fault resistance.

The most critical input to a fault location algorithm is the impedance data. Be sure to use the impedances and voltages and currents appropriate for the type of fault. For line-to-ground faults, use line-to-ground quantities; and for others, use phase-to-phase quantities:

- Line-to-ground fault

$$V = V_a,\ I = I_a,\ Z = Z_S = (2Z_1 + Z_0)/3$$

- Line-to-line, line-to-line-to-ground, or three-phase faults

$$V = V_{ab},\ I = I_a - I_b,\ Z = Z_1$$

Remember that these are all complex quantities. It helps to have software that automatically calculates complex phasors from a waveform. Several methods are available to calculate the rms values from a waveform; a Fourier transform is most common. Some currents have significant offset that can add error to the result. Try to find the magnitudes and angles after the offset has decayed (this is not possible on some faults cleared quickly by fuses).

If potential transformers are connected phase to phase, we can still estimate locations for ground faults if we know the zero-sequence source impedance. Schweitzer (1990) shows that the phase-to-ground voltage is

$$V_a = 1/3(V_{ab} - V_{ca}) - Z_{0,src}I_0$$

where

$Z_{0,src}$ = zero-sequence impedance of the source, Ω

I_0 = zero-sequence current measured during the fault = $I_a/3$ for a single line-to-ground fault on phase A

Although the voltages and currents are complex, we can also estimate the distance just using the absolute values. Although we lose some information on how accurate our solution is because we lose the phase angle information, in many cases it is as good as using the complex quantities. So, the simple fault location solution with absolute values is

$$d = \frac{V}{I \cdot Z_l}$$

where

V = absolute value of the rms voltage during the fault, V

I = absolute value of the rms current during the fault, A

Z_l = absolute value of the line impedance, Ω/length unit

d = distance to the fault, length unit such as mi

With this simple equation, we can estimate answers with voltage and current magnitudes. For a ground fault, Z_l=Z_S is about 1 Ω/mi. If the line-to-ground voltage, V=5000 V, and the fault current, I=1500 A, the fault is at about 3.3 mi (5000/1500). Remember to use the phase-to-phase voltage and $|I_a - I_b|$ (and not $|I_a| - |I_b|$) for faults involving more than one phase.

We can calculate the distance to the fault using only the magnitude of the current (no phase angles needed and only prefault voltage needed) and the line and source impedances involved. If we know the absolute value of the fault current and the prefault voltage and the source impedance, the distance to the fault is a solution to the quadratic equation

$$d = \frac{-b + \sqrt{b^2 - 4ac}}{2a}$$

where

$a = Z_l^2$

$$b = 2R_l R_{src} + 2X_l X_{src}$$

$$c = Z_{src}^2 - \left(\frac{V_{prefault}}{I_{fault}} \right)^2$$

and

R_{src} = source resistance, Ω
X_{src} = source reactance, Ω
Z_{src} = absolute value of the source impedance, Ω
R_l = line resistance, Ω/unit distance
X_l = line reactance, Ω/unit distance
Z_l = absolute value of the line impedance, Ω/unit distance
I_{fault} = absolute value of the rms current during the fault, A
$V_{prefault}$ = absolute value of the rms voltage just prior to the fault, V

In this case, we are doing the same thing as taking a fault current profile (such as Figure 4.11) and interpolating the distance. In fact, it is often much easier to use a fault current profile developed from a computer output rather than this messy set of equations. If the prefault voltage is missing, assume that it is equal to the nominal voltage. If we have the prefault voltage, divide the current by the per-unit prefault voltage before interpolating on the fault current profile. Using a fault current profile also allows changes in line impedances along the length of the line. Carolina Power & Light used this approach, and Lampley (2002) reported that their locations were accurate to within 0.5 mi 75% of the time; and in most of the remaining cases, the fault was usually no more than 1 to 2 mi from the estimate.

We can also just use voltages. If we know the source impedance, we do not need current. The distance calculation is another quadratic formula solution, this time with

$$d = \frac{-b - \sqrt{b^2 - 4ac}}{2a} \qquad \text{(the negative root because } a \text{ is negative)}$$

where

$$a = Z_l^2 - Z_l^2 \left(\frac{V_{prefault}}{V_{fault}} \right)^2$$

$$b = 2R_l R_{src} + 2X_l X_{src}$$

$$c = Z_{src}^2$$

and

V_{fault} = absolute value of the rms voltage during the fault, V

As with the fault current approach, rather than using this equation, we can interpolate a voltage profile graph to find the distance to the fault (such as those in Figure 2.6). Again, we are assuming that the arc impedance is zero.

Fault locations of line-to-line and three-phase faults are most accurate because the ground path is not included. The ground return path has the most uncertainties. The impedance of the ground return path depends on the number of ground rods, the earth resistivity, and the presence of other objects in the return path (cable TV, buried water pipes, etc.). The ground return path is also nonuniform with length.

This type of fault location is useful for approximate locations. For permanent faults, a location estimate helps shorten the lengths of circuit patrolled. Distance estimates can also help find those irksome recurring temporary faults that cause repeated momentary interruptions. Fault locations are most accurate when the fault is within 5 mi of the measurement; beyond that, the voltage profile and fault current profiles flatten out considerably, which increases error. Fault location is difficult if a circuit has many branches. If a fault is 2 mi from the source on phase B, but there are 12 separate circuits that meet that criteria, the location information is not as useful. Fault location is also difficult on circuits with many wire-size changes; it works best on circuits with uniform mainline impedances with relatively short taps. Impedance-based location methods produce close but not pinpoint results. For underground faults, we would like to know exactly where to dig, but these methods do not have that accuracy.

4.3 Limiting Fault Currents

Limiting fault current has many benefits, which improve the safety and reliability of distribution systems:

- *Failures* — Overhead line burndowns are less likely, cable thermal failures are less likely, violent equipment failures are less likely.
- *Equipment ratings* — We can use reclosers and circuit breakers with less interrupting capability and switches and elbows with less momentary and fault close ratings. Lower fault currents reduce the need for current-limiting fuses and for power fuses and allow the use of cutouts and under-oil fuses.
- *Shocks* — Step and touch potentials are less severe during faults.
- *Conductor movement* — Conductors move less during faults (this provides more safety for workers in the vicinity of the line and makes conductor slapping faults less likely).
- *Coordination* — Fuse coordination is easier. Fuse saving is more likely to work.

At most distribution substations, three-phase fault currents are limited to less than 10 kA, with many sites achieving limits of 7 to 8 kA. The two main ways that utilities manage fault currents are:

- *Transformer impedance* — Specifying a higher-impedance substation transformer limits the fault current. Normal transformer impedances are around 8%, but utilities can specify impedances as high as 20% to reduce fault currents.
- *Split substation bus* — Most distribution substations have an open tie between substation buses, mainly to reduce fault currents (by a factor of two).

Line reactors and a neutral reactor on the substation transformer are two more options used to limit fault currents, especially in large urban stations where fault currents may exceed 40 kA.

There are drawbacks of increasing impedance to reduce fault currents. Higher impedance reduces the stiffness of the system: voltage sags are worse, voltage flicker is worse, harmonics are worse, voltage regulation is more difficult.

A reactor in the substation transformer neutral limits ground fault currents. Even though the neutral reactor provides no help for phase-to-phase or three-phase faults, it provides many of the benefits of other methods of fault reduction. Neutral reactors cost much less than line reactors. Ground faults are the most common fault; and for many types of single-phase equipment, the phase-to-ground fault is the only possible failure mode. A neutral reactor does not cause losses or degrade voltage regulation to the degree of phase reactors. On the downside, a neutral reactor has a cost and uses substation space, and a neutral reactor reduces the effectiveness of the grounding system.

Several advanced fault-current limiting devices have been designed (EPRI EL-6903, 1990). Most use some sort of nonlinear elements — arresters, saturating reactors, superconducting elements, or power electronics such as a gate-turn-off thyristor — to limit the fault current either through the physics of the device or through computer control. Since most distribution systems have managed fault currents sufficiently well, these devices have not found a market. Given that, the EPRI study surveyed utilities and found evidence that a market for fault-current limiters exists if a device had low enough cost and was robust enough.

4.4 Arc Characteristics

Many distribution faults involve arcs through the air, either directly through the air or across the surface of hardware. Although a relatively good conductor, the arc is a very hot, explosive fireball that can cause further damage at the fault location (including fires, wire burndowns, and equipment damage). This section discusses some of the physical properties of arcs, along with the ways in which arcs can cause damage.

Normally, the air is a relatively good insulator, but when heavily ionized, the air becomes a low-resistance conductor. An arc stream in the air consists of highly ionized gas particles. The arc ionization is due to *thermal* ionization caused by collisions from the random velocities of particles (between electrons, photons, atoms, or molecules). Thermal ionization increases with increasing temperature and with increasing pressure. The heat produced by the current flow (I^2R) maintains the ionization. The arc stream has very low resistance because there is an abundance of free, charged particles, so current flow can be maintained with little electric field. Another type of ionization caused by acceleration of electrons from the electric field may initially start the ionization during the electric-field breakdown, but once the arc is created, electric-field ionization plays a less significant role than thermal ionization.

One of the characteristics that is useful for estimating arc-related phenomenon is the arc voltage. The voltage across an arc remains constant over a wide range of currents and arc lengths, so the arc resistance decreases as the current increases. The voltage across an arc ranges between 25 and 40 V/in (10 to 16 V/cm) over the current range of 100 A to 80 kA (Goda et al., 2000; Strom, 1946). The arc voltage is somewhat chaotic and varies as the arc length changes. More variation exists at lower currents. As an illustration of the energy in an arc, consider a 3-in. (7.6-cm) arc that has a voltage of about 100 V. If the fault current is 10 kA, the power in the arc is $P = V \cdot I = 100\ \text{V} \cdot 10\ \text{kA} = 1\ \text{MW}$. Yes, 1 MW! Arcs are explosive and as hot as the surface of the sun.

An upper bound of roughly 10,000 to 20,000 K on the temperature of the arc maintains the relatively constant arc voltage per unit length. For larger currents, the arc responds by increasing the volume of gas ionized (the arc expands rather than increasing the arc-stream temperature). Higher currents increase the cross-sectional area of the arc, which reduces the resistance of the arc column; the current density is the same, but the area is larger. So, the voltage drop along the arc stream remains roughly constant. The arc voltage depends on the type of gas and the pressure. One of the reasons an arc voltage under oil has a higher voltage gradient than an arc in air is because the ionizing gas is mainly hydrogen, which has a high heat conductivity. A high heat conductivity causes the arc to restrict and creates a higher-density current flow (and more resistance). The arc voltage gradient is also a function of pressure. For arcs in nitrogen (the main ionizing gas of arcs in air), the arc voltage increases with pressure as $V \propto P^k$, where k is approximately 0.3 (Cobine, 1941).

Another parameter of interest is the arc resistance. A 3-ft (1-m) arc has a voltage of about 1400 V. If the fault current at that point in the line was 1000 A, then the arc resistance is about 1.4 Ω. A 1-ft (0.3-m) arc with the same fault current has a resistance of 0.47 Ω. Most fault arcs have resistances of 0

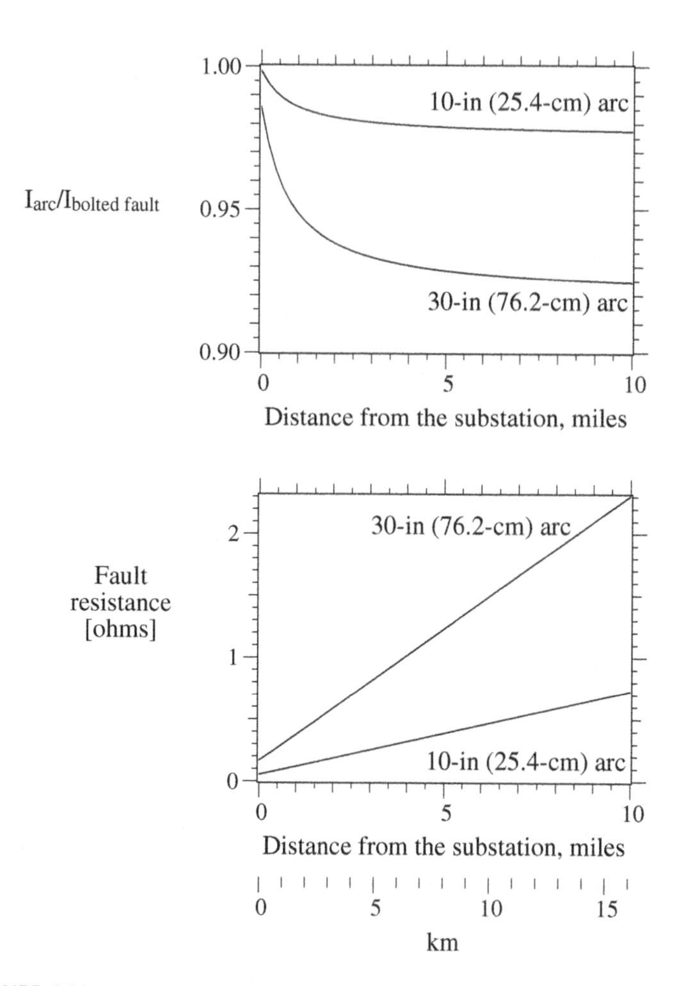

FIGURE 4.21
Ratio of fault current with and without an arc of the given length on a 12.47-kV circuit. This assumes the same system parameters as the fault profile in Figure 4.11 with the following additional assumptions: the arc voltage gradient equals 40 V/in. (16 V/cm), the arc voltage is all resistive, and the nonlinearity of the arc voltage is ignored.

to 2 Ω. Figure 4.21 shows that the impact of an arc on the fault currents along the line is fairly minor.

An arc voltage waveform has distinguishing characteristics. Figure 4.22 shows an arcing fault voltage that was initiated by tree contact on a 13-kV circuit measured during the EPRI DPQ project. The voltage on the arc is in phase with the fault current (it is primarily resistive). When the arc current goes to zero, the arc will extinguish. The recovery voltage builds up quickly because of the stored energy in the system inductance. Voltage builds to a point where it causes arc reignition. The reason for the blip at the start of the waveform (it is not a straight square wave) is that the arc cools off at the current zero. Cooling lowers the ionization rate and increases the arc resis-

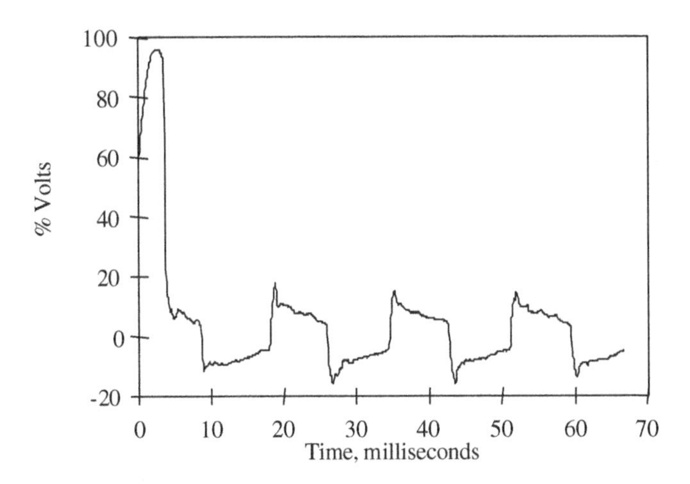

FIGURE 4.22
Arcing fault measured during the EPRI DPQ study. (Copyright © 1996. Electric Power Research Institute. TR-106294-V3. *An Assessment of Distribution System Power Quality: Volume 3: Library of Distribution System Power Quality Monitoring Case Studies.* Reprinted with permission.)

tance. Once it heats up again, the voltage characteristic flattens out. The waveform is high in the odd harmonics and for many purposes can be approximated as a square wave.

The movement and growth of an arc is primarily in the vertical direction. Tests at IREQ in Quebec showed that the primary reason that the arc elongates and moves vertically is the rising hot gases of the arc (Drouet and Nadeau, 1979). The magnetic forces ($J \times B$) did not dominate the direction or elongation. As a first approximation over a range of currents between 1 and 20 kA, arc voltages up to 18 kV, and durations up to 0.5 sec, the arc length can be expressed as a function of the duration only as

$$l = 30t$$

where
l = arc length, m
t = fault duration, sec

The arc movement is a consideration for underbuilt distribution and for vertical construction. The equation above can be used as an approximation to determine if an underbuilt distribution circuit could evolve and fault a distribution or transmission circuit above. It also gives some idea of how faults can evolve to more than one phase. Figure 4.23 shows an example of a fault evolving from one to three phases over the course of about 1.5 sec (the construction type is unknown, but it is probably a horizontal configuration). Given the vertical movement of a fault current arc, vertical designs are more prone to having faults evolve to more than one phase. We might

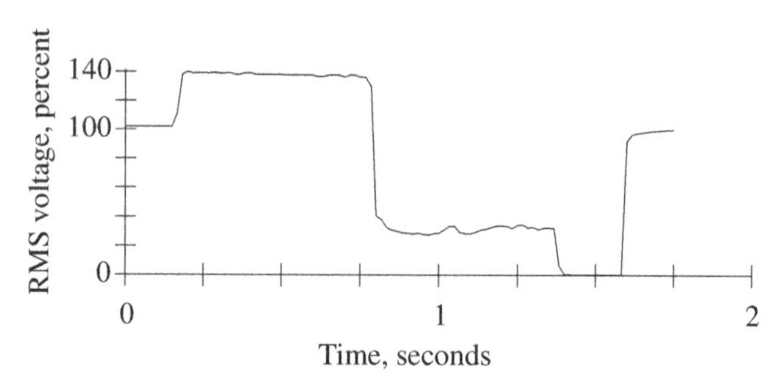

FIGURE 4.23
Voltage waveform from a fault that started as a single-phase fault (indicated by a swell on the phase shown), evolved to a double-line-to-ground fault (the voltage sags to about 35%), and finally evolved to a three-phase-to-ground fault (where the voltage sags to zero). (Copyright © 1996. Electric Power Research Institute. TR-106294-V3. *An Assessment of Distribution System Power Quality: Volume 3: Library of Distribution System Power Quality Monitoring Case Studies.* Reprinted with permission.)

think that a fault evolving to include other phases is not a concern since the three-phase circuit has to be opened anyway, whether it is a single- or three-phase fault. But, the voltage sag during the fault is more severe if more than one phase is involved, which is a good reason to use designs that do not tend to propagate to more than one phase (and to use relaying or fusing that operates quickly enough to prevent it from happening).

The temperature in the arc can be on the order of 10,000 K. This heat creates hazards from burning and from the pressure wave developed during the fault. The longer the arc, the more energy is created. NFPA provides guidelines on safe distances for workers based on arc blasts (NFPA 70E-2000). Several groups have worked to determine the appropriate characteristics of protective clothing (ASTM F1506-94, 1994). Because of the pressure wave, consider hearing protection and fall protection for workers who could be exposed to fault arcs. Arcs can cause fires: pole fires or fires in oil-filled equipment such transformers.

The pressure wave from an arc in an enclosed substation was the probable cause of a collapse of a substation building (important since many distribution stations are required to be indoors because of environmental considerations). Researchers found during tests that the pressure from a fault arc can be approximated (Drouet and Nadeau, 1979) by

$$A = 1.5\frac{I \cdot t}{l}$$

where

A = pressure, kN/m^2 (1 kN/m^2 = 20.9 lb/ft^2)
I = fault current magnitude, kA
t = duration of the fault, sec
l = distance from the source, m (for $l > 1$ m)

Although many electrical damage characteristics are a function of $\int I^2\, dt$, the pressure wave is primarily a function of $\int I\, dt$ (because the voltage along the arc length is constant and relatively independent of the arc current). Where arcs attach to wires, melting weakens wires and can lead to wire burndown. Most tests have shown that the damage is proportional to $\int I^k\, dt$, where k is near one but varies depending on the conductor type. For burndowns or other situations where the arc burns the conductor, the total length of the arc is unimportant, the small portion of the arc near the attachment point is important. The voltage drop near the attachment point is also very constant and does not vary significantly with current. The damage to conductors is very much like that of an electrical arc cutting torch.

Burndowns are much more likely on covered wire (also called tree wire). The covering restricts the movement of the attachment point of the arc to the conductor. On bare wire, the arc will move because of the heating forces on the arc and the magnetic forces (also called motoring).

On bare wire, burndowns are a consideration only on small conductors. Tests (Lasseter, 1956) have shown that the main cause of failure on small aluminum conductors is that the hot gases from the arc anneal the aluminum, which reduces tensile strength. The testers found little evidence of arc burns on the conductors. Failures can occur midspan or at a pole. Motoring is not fast enough to protect the small wire.

Arcs can damage insulators following flashover along the surface of the insulators. This was the primary reason for the development of arcing horns for transmission-line insulators. Arcing horns encourage a flashover away from the insulator rather than along the surface. Arcs can fail distribution insulators. During fault tests across insulators by Florida Power & Light (Lasseter, 1965), the top of the arc moved along the conductor. The point of failure was at the bottom of the insulator where the arc moved up the pin to the bottom edge of the porcelain. The bottom of the insulator gets very hot and can fail from thermal shock. The threshold of chipping was about 360 C (C = coulombs = A-sec = $\int I\, dt$), and the threshold of shattering was about 1125 C (see Figure 4.24). Adding an aluminum or copper washer (but not a steel washer) on top of the crossarm under the flange of the grounded steel pin reduced insulator shattering. The arc attaches to the washer rather than moving up along the pin, increasing the threshold of chipping by a factor of five. Composite insulators perform better for surface arcs than porcelain insulators (Mazurek et al., 2000). Some composite insulators have an external arc withstand test where $I{\cdot}t$ shall be 150 kA-cycles (2500 C for 60 Hz) (IEEE Std. 1024-1988).

Distribution voltages can sustain very long arcs, but self-clearing faults can occur such as when a conductor breaks and falls to the ground (stretching an arc as it falls). The maximum arc length is important because the longer the arc, the more energy is in the arc. For circuits with fault currents on the order of 1000 A and where the transient rise to the open-circuit voltage is about 10 µs, about 50 V may be interrupted per centimeter of arc length (Slepian, 1930). For a line-to-ground voltage of 7200 V, a line-to-ground arc

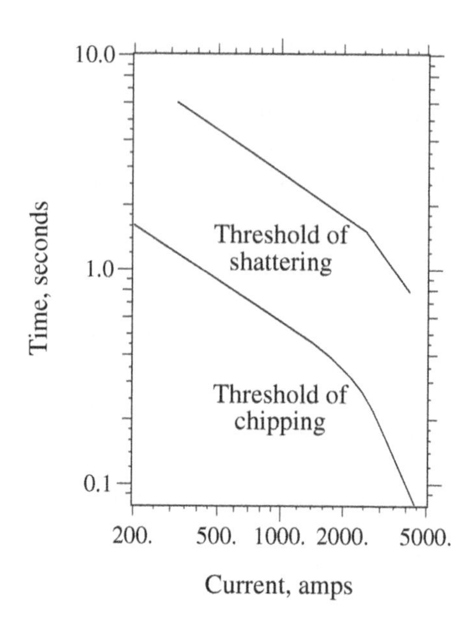

FIGURE 4.24
Insulator damage characteristics. (Data from [Lasseter, 1965].)

can reach a length of about 12 ft (3.7 m) before it clears. As another approximation, the length that an arc can maintain in resistive circuits is [from (Rizk and Nguyen, 1984) with some reformulation]:

$$l = V\sqrt{I}$$

where

l = arc length, in. (1 in. = 2.54 cm)
I = rms current in the previous half cycle, A
V = system rms voltage, kV (line to ground or line to line depending on fault type)

4.5 High-Impedance Faults

When a conductor comes in physical contact with the ground but does not draw enough current to operate typical protective devices, you have a *high-impedance* fault. In the most common scenario, an overhead wire breaks and falls to the ground (a *downed wire*). If the phase wire misses the grounded neutral or another ground as it falls, the circuit path is completed by the high-impedance path provided by the contact surface and the earth.

The return path for a conductor lying on the ground can be a high impedance. The resistance varies depending on the surface of the ground. Table

TABLE 4.3

Typical High-Impedance Fault Current Magnitudes

Surface	Current, A
Dry asphalt	0
Concrete (no rebar)	0
Dry sand	0
Wet sand	15
Dry sod	20
Dry grass	25
Wet sod	40
Wet grass	50
Concrete (with rebar)	75

Source: IEEE Tutorial Course 90EH0310-3-PWR, "Detection of Downed Conductors on Utility Distribution Systems," 1990. With permission. ©1990 IEEE.

4.3 shows typical current values measured for conductors on different surfaces (for 15-kV class circuits).

The frequency of high-impedance faults is uncertain. Most utilities responding to an IEEE survey reported that high-impedance faults made up less than 2% of faults while a sizeable number (15% of those surveyed) suggested that between 2 and 5% of distribution faults were not detectable (IEEE Working Group on Distribution Protection, 1995). Even with small numbers, high-impedance faults pose an important safety hazard.

On distribution circuits, high-impedance faults are still an unsolved problem. It is not for lack of effort; considerable research has been done to find ways to detect high-impedance faults, and progress has been made [see (IEEE Tutorial Course 90EH0310-3-PWR, 1990) for a more in-depth summary]. Research has identified many characteristics of high-impedance faults and have tested them for detection purposes. Efforts have been concentrated on detection at the substation based on phase and ground currents. High-impedance faults usually involve arcing, and arcing generally creates the lower odd harmonics. Arcing faults may also contain significant 2- to 10-kHz components. Arcing also bursts in characteristic patterns. High-impedance faults often cause characteristic changes in the load (for example, a broken conductor will drop the load on that phase). None of these detection methods is perfect, so some detection schemes use more than one method to try to detect high-impedance faults.

We can also detect broken conductors at the ends of radial circuits. Loss of voltage is the simplest method. Communication to an upstream protective device or to a control center is required. A difficulty is that it takes many devices to adequately cover a radial circuit (depending on how many branches occur on the circuit). Also, the "ends" of circuits could change during circuit reconfigurations or sectionalizing due to circuit interruptions. Also, if the loss-of-voltage detector is downstream of a fuse, another detector is needed at the fuse, so we can determine if the fuse operated or the conductor broke.

Practices that help reduce high-impedance faults include

- *Tight construction framings* — If a phase wire breaks, it is more likely to contact a neutral as it falls. (A drawback is that utilities have reported poorer reliability with tighter constructions like the armless design.) A vertical construction is better than a horizontal construction. Single-phase structures are better than three-phase structures.
- *Stronger conductors* — Larger conductors or ACSR instead of all-aluminum conductor are stronger and less likely to break for a given mechanical or arcing condition.
- *Smaller/faster fuses* — Faster fuses are more likely to operate for high-impedance faults. In addition, small fuses are likely to clear before arcing damages wires, which could burn down the wires.
- *Tree trimming* — Clearing trees and trimming reduces the number of trees or branches breaking conductors.
- *Fewer reclose attempts* — Each reclose attempt causes more damage at the fault location.
- *Higher primary voltages* — High-impedance faults are much less likely at 34.5 kV and somewhat less likely at 24.94 kV than 15-kV class voltages.
- *Public education* — Public advertisements warning the public to stay away from downed wires help reduce accidents when high-impedance faults do occur.

Practices to *avoid* include:

- *Covered wire* — Burndowns are more likely with covered wire. If a covered wire does contact the ground, it is less likely to show visible signs that it is energized such as arcing or jumping which would help keep bystanders away.
- *Unfused taps* — Burndowns are more likely with the smaller wire used on lateral taps.
- *Midspan connectors* — Flying taps can cause localized heating and mechanical stress.
- *Rear-lot construction* — Rear-lot construction is not as well maintained as road-side construction, so trees are more likely to break wires. If wires do come down, it is more hazardous since they are coming down in someone's backyard.
- *Neutral on the crossarm* — If a phase wire breaks, it is much less likely to contact the neutral as it falls if the neutral is on the crossarm. Other constructions that may have this same problem are overhead shield wires and spacer cables that do not have an additional neutral below.

Three-wire distribution systems have some advantages and some disadvantages related to high-impedance faults. The main advantage of three-wire systems is that there is no unbalanced load. A sensitive ground relay can be used, which would detect many high-impedance faults. The sensitivity of the ground relay is limited by the line capacitance. The main disadvantage of three-wire systems is that there is no multigrounded neutral. If a phase conductor breaks, there is a high probability that there will be a high-impedance fault. If there is underbuilt secondary or phone or cable TV under the three-wire system, then a high-impedance fault is less likely because a grounded conductor is below the phases.

Spacer cable has some mechanical strength advantages that could help keep phase wires in the air, and it has fewer faults due to trees. A downside is the covering which makes burndowns more likely. Also, it has a messenger wire that may act as the neutral; if it does not also have an underbuilt neutral, a phase conductor is more likely to fall unimpeded.

Backfeeds from three-phase transformer and capacitor installations can cause dangerous situations. If a wire breaks near a pole, at least half of the time, the load side (downstream side) will lie on the ground. Backfeed to the downed wire can occur through three-phase transformers downstream of the fault. The backfeed can provide enough voltage and current to the downed wire to be dangerous (but there will not be enough current to trip protective devices). Note that a grounded-wye – grounded-wye connection does not eliminate backfeeds. Another backfeed scenario is shown in Figure 4.25 where a three-phase load is fed from a fused tap. A bolted, low-impedance fault on one phase will blow the fuse, but backfeed current may flow through the three-phase connection. Even with a low-impedance fault, the backfeed current can be low enough that neither the remaining tap fuses nor the transformer fuses will operate. The fault can continue to arc until the wire burns down. An ungrounded capacitor bank can also provide a backfeed path.

Commercial relays that have high-impedance fault detection capabilities are available. One of the main problems with detection of high-impedance faults is false fault detections. If a detection system in the station detects a fault and a whole feeder is tripped for an event that is not really a high-impedance fault, reliability is severely hurt. Before reenergizing, crews patrol the circuit and make sure there was not a downed conductor. The sensitivity of a device must be traded off against its dependability. If it is too sensitive, many false operations will result. An alternative to tripping is alarming. Operators in control centers may always trip a circuit if it signals a high-impedance fault for fear of discipline if it really was a high-impedance fault and an accident occurred. Each time a high-impedance fault is detected, crews would have to patrol the circuit. If operators have too many false alarms, they may ignore the alarm.

A tree branch in contact with a phase conductor also forms a high-impedance fault. This is not as dangerous as a downed wire. Most of the time the circuit operates normally, without danger to the public. A tree branch in

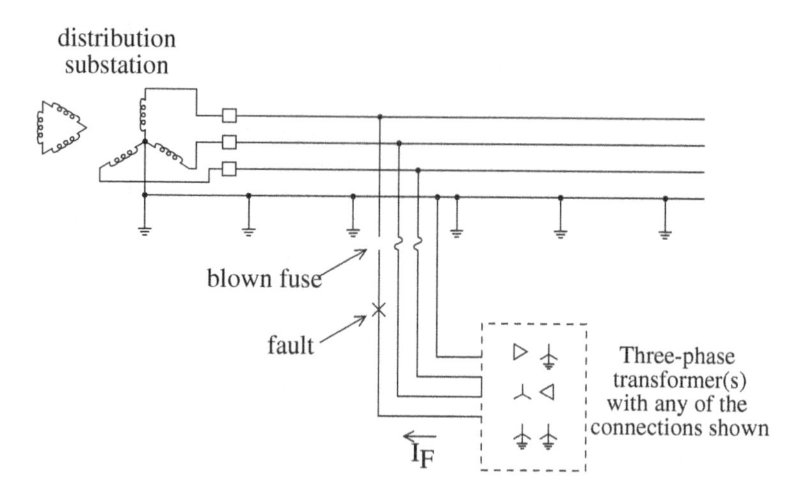

FIGURE 4.25
Backfeed to a fault downstream of a blown fuse.

contact with a phase conductor can draw enough current to trip a high-impedance fault detector. In a heavily treed area, crews could require many hours to find the location where the tree contact is taking place.

One of the main problems with substation detection is that each feeder usually covers many miles of line. With most faults, lateral fuses provide an effective way to isolate and identify the location of faults within a relatively small area. While it would be nice to have high-impedance fault detection capability on lateral taps, the costs have been prohibitive. Contrast a station detection scheme (one device) with detectors at taps — tens or hundreds of devices.

Another solution to falling conductors is using guards installed on poles below the phase wires to "catch" phase conductors. The guards are connected to the grounded neutral, so when a phase conductor breaks, a low-impedance fault is created. This would be a significant expense to install system-wide, but it may be suitable for isolated locations where it is critical not to have energized downed conductors (a stretch that runs across a school playground or a span that crosses a major road).

4.6 External Fault Causes

4.6.1 Trees

For many utilities, trees cause more faults and more interruptions than any other factor. Tree trimming is expensive — the largest maintenance item for most distribution companies. Tree trimming is also a contentious issue with the public. Residents hate to have their 100-year-old maple trees touched (or even their 30-year-old cottonwoods).

Faults caused by trees generally occur from three conditions:

1. Falling trees knock down poles or break pole hardware.
2. Tree branches blown by the wind push conductors together.
3. A branch falls across the wires and forms a bridge from conductor to conductor (or natural tree growth causes a bridge).

Tree-caused faults can be temporary or permanent. Falling trees or branches can cause permanent faults. Either falling across wires or pushing them together, tree branches can cause temporary faults. Broken tree branches account for the majority of interruptions. In one utility in the northeast U.S., 63% were caused by broken branches compared to 11% from falling trees and only 2% from tree growth (Simpson, 1997). Niagara Mohawk Corporation (NIMO) found that 86% of permanent tree-faults were outside of the right-of-way, and most were from major breakage (see Figure 4.26). Falling trees do the most damage; they often break conductors and even poles in some cases. Tree faults usually occur during storms, primarily from wind. Snow and ice additionally contribute to tree failures.

Several companies have done tests to evaluate the electrical properties of tree branches and how they cause faults (Goodfellow, 2000; Williams, 1999). For a tree branch to cause a fault, the branch must bridge the gap between two conductors, which usually must be sustained for more than 1 min. A tree touching just one conductor will *not* fault. The tree branch must cause a connection between two bare conductors (it can be phase to phase or phase to neutral). A tree branch into one phase conductor normally draws less than one amp of current under most conditions; this may burn some leaves, but it will not fault. On small wires in contact with a tree, the arcing to the tree may be enough to burn the wire down under the right conditions. While the tree in contact with one wire will not fault the circuit, there are some safety issues with trees in contact with overhead conductors.

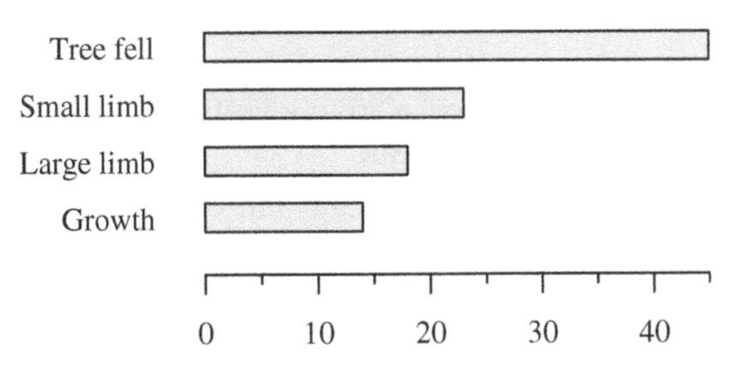

FIGURE 4.26
Tree failure causes for the Niagara Mohawk Power Corporation. (Data from [Finch, 2001].)

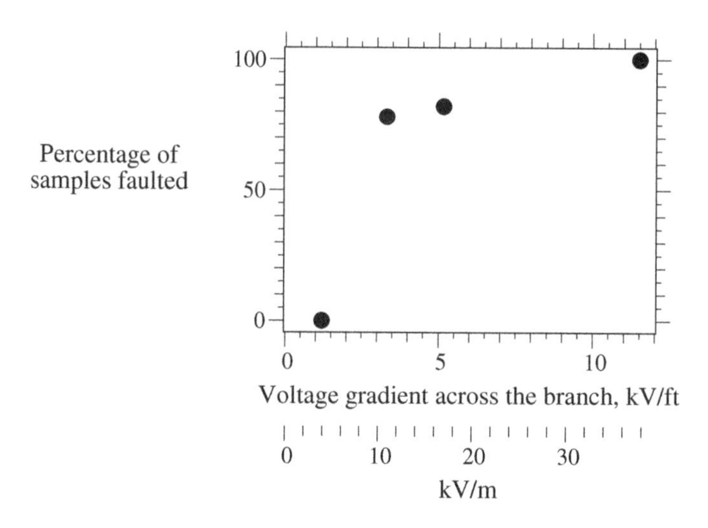

FIGURE 4.27
Percentage of samples faulted based on the voltage gradient across the tree branch. (Data from [Goodfellow, 2000].)

A fault across a tree branch between two conductors takes some time to develop. If a branch falls across two conductors, arcing occurs at each end where the wire is in contact with the branch. At this point in the process, the current is small (the tree branch has a relatively high impedance). The arcing burns the branch and creates carbon by oxidizing organic compounds. The carbon provides a good conducting path. Arcing then occurs from the carbon to the unburned portion of the branch. A carbon track develops at each end and moves inward.

Once the carbon path is established completely across the branch, the fault is a low-impedance path. Now the current is high; it is effectively a bolted fault. It is also a permanent fault. If a circuit breaker or recloser is opened and then reclosed, the low-impedance carbon path will still be there unless the branch burns enough to fall off of the wires.

Some other notable electrical effects include the following:

- It makes little difference if the branch is wet or dry. Live branches are more likely to fault for a given voltage gradient, but dead branches are more likely to break.
- Little branches can burn through and fall off before the full carbon track develops, so minor leaf and branch burning does not cause faults.
- The likelihood of a fault depends on the voltage gradient along the branch (see Figure 4.27).
- The time it takes for a fault to occur also depends on the voltage gradient (see Figure 4.28).
- Lower voltage circuits are much more immune to flashovers from branches across conductors. A 4.8-kV circuit on a 10-ft crossarm has

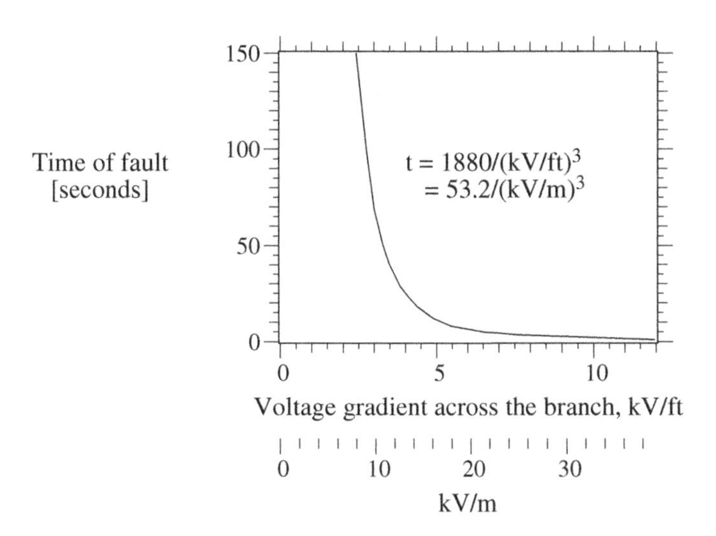

FIGURE 4.28
Time to fault based on the voltage gradient across the tree branch. (Data from [Goodfellow, 2000] with the curvefit added)

about a phase-to-phase voltage gradient of 1 kV/ft, very unlikely to fault from tree contact. A 12.47-kV circuit has a 2.7 kV/ft gradient, which is more likely to fault.

- Candlestick or armless designs are more likely to flashover because of tighter conductor-to-conductor spacings.

These effects reveal some key issues:

- Trimming around the conductors in areas with a heavy canopy does not prevent tree faults. Traditionally, crews trim a "hole" around the conductors with about a 10-ft (3-m) radius. If there is a heavy canopy of trees above the conductors, this trimming strategy performs poorly since most faults are caused by branches falling from above.
- Vertical construction may help since the likelihood of a phase-to-phase contact by falling branches is reduced.
- Three-phase construction is more at risk than single-phase construction.

Tree trimming is expensive. An EPRI survey found that utilities spend an average of about $10 per customer each year on tree trimming (EPRI TR-109178, 1998). Trimming can also irritate communities. It is always a dilemma that people do not want their trees trimmed, but they also do not want interruptions. Consider the following general tree-trimming guidelines:

- *Removal* — This is the most effective fault-prevention strategy, and many homeowners are willing to have trees removed.

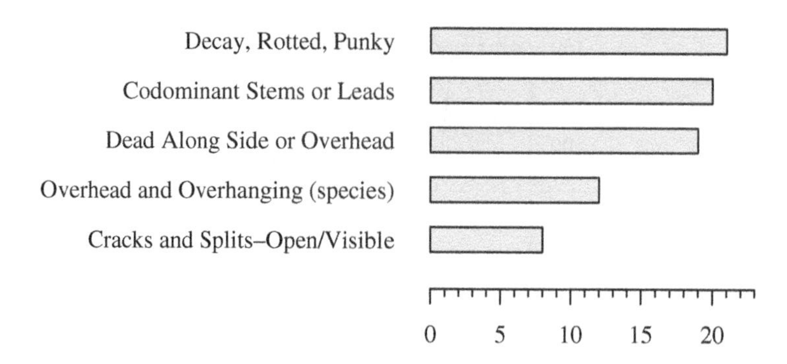

FIGURE 4.29
Defects causing tree failure for the Niagara Mohawk Power Corporation. (Data from [Finch, 2001].)

- *Danger trees* — Trimming/removal is most effective if trees and branches that are likely to fail are removed or trimmed to safe distances. This does take some expertise by tree trimming crews.
- *Target* — As with any fault-reduction program, efforts are best spent on the poorest performing circuits that affect the most customers. Along the same thought, spend more on three-phase mains than on single-phase taps.

Targeting danger trees is especially beneficial but requires expertise. In a careful examination of several cases where broken branches or trees damaged the system, 64% of the trees were living (Finch, 2001). Finch also advises examining trees from the backside, inside the tree line (defects on that side are more likely to fail the tree into the line). Finch describes several defects that help signal danger trees (see Figure 4.29). Dead trees or large splits are easy to spot. Cankers (a fungal disease) or codominant stems (two stems, neither of which dominates, each stem at a branching point is approximately the same size) require more training and experience to detect. It also helps to know the types of trees that are prone to interruptions (this varies by area and types of trees). For Niagara Mohawk, black locusts and aspens are particularly troublesome; large, old roadside maples also caused more than their share of damage (see Table 4.4).

Acceptable tree trimming (that is also still effective) is a public relations battle. Some strategies that help along these lines include:

- Talking to residents prior to/during tree trimming.
- Trimming trees during the winter (or tree-trimming done "under the radar") — The community will not notice tree trimming as much when the leaves are not on the trees.

TABLE 4.4

Comparison of Trees Causing Permanent Faults with the Tree Population

Species	Percent of Outages	Percent of New York State Population
Ash	8	7.9
Aspen	9	0.6
Black Locust	11	0.3
Black Walnut	5	N/A
Red Maple	14	14.7
Silver Maple	5	0.2
Sugar Maple	20	12.0
White Pine	6	3.3

Source: Finch, K., "Understanding Tree Outages," EEI Vegetation Managers Meeting, Palm Springs, CA, May 1, 2001.

- Trimming trees during storm cleanups. Right after outages, residents are more willing to accept their beloved trees being hacked up (this is a form of the often practiced "storm-induced maintenance"; fix it when it falls down).
- Cleaning up after trees are trimmed/removed.
- Offering free firewood.

Choosing a tree-trimming cycle is tricky. Many utilities use a three- to five-year cycle. Longer tree-trimming cycles lead to higher fault rates (see Figure 4.30). The optimal trimming cycle depends on

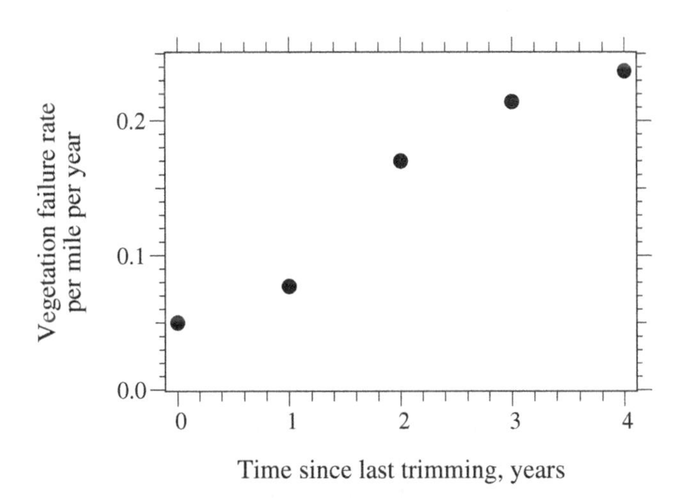

FIGURE 4.30
Tree failure rates vs. time since last trimming for one Midwestern utility. (Data from [Kuntz, 1999].)

TABLE 4.5

Interruption Rates in Outages per 100 Miles per Year Comparing Bare Wire, Tree Wire, and Spacer Cable at One Utility in the U.S. from 1995 to 1997

Fault Type	Bare	Tree Wire	Spacer Cable
Tree related	17.6	6.6	1.8
Animals	12.1	5.9	2.9
Lightning	3.4	1.9	1.0
Unknown	5.9	2.6	1.0
All other	11.3	4.6	5.9
Totals	50.3	21.7	12.5

Source: Hendrix, "Reliability of Overhead Distribution Circuits," Hendrix Wire & Cable, Inc., August 1998.

- Type of trees, growth rates, and growing conditions
- Community tolerance for trimming
- Economic assumptions, especially the chosen time value of money

In heavily treed areas, covered conductors help reduce tree faults. This "tree wire" provides extra insulation that reduces the chance of flashover for a branch between conductors. Good fault data is hard to find comparing fault rates of bare wire with covered wire. One utility whose results are provided in Table 4.5 has shown reductions in interruption rates of greater than 50% for covered wire and even more for spacer cable (the only caveat here is the data is published by a manufacturer, so it may not be unbiased). Tree and animal faults were also reduced by over 50%. European experience with covered conductors suggests that covered-wire fault rates are about 75% less than bare-wire fault rates. In Finland, fault rates on bare lines are about 3 per 100 km/year on bare and 1 per 100 km/year on covered wire (Hart, 1994). Covered wire helps with animal faults as well as tree faults.

Spacer cable and aerial cables are also alternatives that perform well in treed areas. Spacer cables are a bundled configuration using a messenger wire holding up three-phase wires that use covered wire. Aerial cables have fully-rated insulation just like underground cables. In South America, both covered wire and a form of aerial cable have been successfully used in treed areas (Bernis and de Minas Gerais, 2001). The Brazilian company CEMIG found that spacer cable faults were lower than bare-wire circuits by a 10 to 1 ratio (although the article did not specify if this included both temporary and permanent faults). The aerial cable faults were lower than bare wire by a 20 to 1 ratio. The effect on interruption durations is shown in Table 4.6. Several spacer cables or aerial cables can be constructed on a pole. Spacer cables and aerial cables have some of the same burndown considerations as covered wire. Spacer cable construction does have a reputation for being hard to work with. Both spacer cable and aerial cable costs more than bare

TABLE 4.6

Comparison of the Reliability Index SAIDI (Average Hours of Interruption per Customer per Year) of Bare Wire, Spacer Cable, and Aerial Cable in Brazil

Construction	SAIDI, h
Bare wire	9.9
Spacer cable	4.7
Aerial cable	3.0

Source: Bernis, R. A. O. and de Minas Gerais, C. E., "CEMIG Addresses Urban Dilemma," *Transmission and Distribution World*, vol. 53, no. 3, pp. 56–61, March 2001.

wire. CEMIG estimated that the initial investment was returned by the reduction in tree trimming. They did minimal trimming around aerial cable (an estimated factor of 12 reduction in maintenance costs) and only minor trimming around spacer cable (an estimated factor of 6 reduction in maintenance costs).

4.6.2 Weather and Lightning

Many faults on overhead circuits are weather related: icing, wind, and lightning. The fault rate during severe storms increases dramatically. Much of the physical and electrical stresses from these events are well beyond the design capability of distribution circuits.

Overhead circuits are designed to NESC (IEEE C2-1997) mechanical standards and clearances, which prescribe the performance of the line itself to the normal severe weather that the poles and wires and other structures must withstand. Most storm failures are from external causes, usually wind blowing tree limbs or whole trees into wires. These cause faults and can bring down whole structures.

Lightning causes many faults on distribution circuits. While most are temporary and do not do any damage, 5 to 10% of lightning faults permanently damage equipment: transformers, arresters, cables, insulators. Distribution circuits do not have any direct protection against lightning-caused faults since distribution insulation cannot withstand lightning voltages. If lightning hits a line, it causes a fault nearly 100% of the time. Since most lightning-caused faults do not do any permanent damage, reclosing is used to minimize the impact on customers. After the circuit flashes over (and there's a fault), a recloser or reclosing circuit breaker will open and, after a short delay, reclose the circuit.

It is important to properly protect equipment from lightning. Transformers and cables are almost always protected with surge arresters. This prevents most permanent faults caused by lightning.

4.6.3 Animals

Faults caused by animals are often the number two cause of outages for utilities (after trees). Squirrels cause the most faults. Squirrels thrive in suburbs and love trees; utilities have noted increases in squirrel faults following development of wooded areas. Squirrels are usually active in the morning and sleep at night. Squirrel faults usually occur in fair weather. The patterns of animal-caused faults have been used to classify "unknown" faults (Mo and Taylor, 1995).

The two main ways to protect equipment against animals (particularly squirrels) are:

- Bushing protectors
- Covered lead wires

Both of these were rated "very good" at reducing animal-caused interruptions in an EPRI survey (EPRI TE-114915, 1999). Several survey respondents noted that the bushing protectors were susceptible to deterioration and tracking (they rated only "good" for durability). Some of the other comments regarding bushing protectors include:

- Insects nest in the bushing coverings, and birds probing for insects cause bird electrocutions and faults.
- Bushing covers hide loose connections and insulator damage and interfere with infrared inspections.

Bushing protectors and covered lead wires are inexpensive if installed with equipment (but expensive to retrofit). For transformer bushing protectors, have crews leave some room between the bottom of the bushing protector and the tank, so water does not build up and leak down through the bushing. Some additional items that also help include

- Trimming trees — Squirrels get to utility equipment via trees (pole climbing is less common). If trees are kept away from lines, utility equipment is less attractive.
- Good outage tracking — Many outages are repeated, so a good outage tracking system can pinpoint hotspots to identify where to target maintenance.
- Identifying animal — If outages are tracked by animal, it is easier to identify proper solutions.
- Maintaining proper clearances
- Avoiding metal crossarms

Animal faults vary by construction habits within a region. Some common problem areas that can lead to frequent animal faults include

- *Transformer bushings* — A very common animal fault is across a transformer bushing. Insulating paints are available for transformers, but it degrades quickly. Bushing guards and/or insulated lead wires offer the best protection.
- *Arresters* (especially polymer) — Another common animal fault is across an arrester, especially a tank-mounted arrester. Polymer-housed arresters have more problems than porcelain-housed arresters because they are much shorter. Use animal guards on tank-mounted arresters (especially on polymer-housed arresters).
- *Cutouts* — Cutouts are sometimes installed such that there is a low clearance between a phase conductor and a grounded object.

Fusing can also change the impact of animal-caused faults for faults across a distribution transformer bushing or a tank-mounted arrester. If the transformer is externally fused, only the customers on the transformer have an outage. If the transformer is a CSP with an internal fuse, then the tap fuse or upstream circuit breaker operates.

Birds rank second (behind squirrels) as far as the number of outages caused by animals (EPRI TE-114915, 1999; Frazier and Bonham, 1996). Many of the practices listed above can help with birds as well. Additionally, some bird-specific practices include:

- Get rid of nests.
- Track as a separate category.
- Remove nearby roosting areas.

The types of animals causing faults varies considerably by region, and there is also significant variation within a region. Animal faults also ebb and flow with animal populations. Animal population data can be used as one way to determine if "unknown" faults are really being caused by certain animals.

4.6.4 Other External Causes

Automobiles and poles do not always coexist nicely. It is difficult for utilities to prevent car accidents. Sometimes utilities can work with city engineers and police to try to lower speed limits and manage traffic better to avoid bad spots. Reflectors on poles may help. Siting poles further from roads also helps.

Balloons and other debris also cause many interruptions. Covered wire helps. Ladders, cranes, and other tall equipment into primary lines cause dangerous ground-level voltages as well as causing faults. Keep getting the word out. Public awareness campaigns help.

TABLE 4.7

Permanent-Fault Causes

Source	Rural	Urban
Equipment failures	14.1%	18.4%
Loss of supply	7.8%	9.6%
External factors	78.1%	72.0%

Source: Horton, W. F., Golberg, S., and Volkmann, C. A., "The Failure Rates of Overhead Distribution System Components," IEEE Power Engineering Society Transmission and Distribution Conference, 1991. With permission. ©1991 IEEE.

4.7 Equipment Faults

Equipment failures — transformers, capacitors, splices, terminations, insulators, connectors — cause faults. When equipment fails, it is almost always as a short circuit and rarely as an open circuit.

Equipment failures on overhead circuits are usually a small percentage of faults (see Figure 4.1). This is confirmed by another study at Pacific Gas & Electric Co. shown in Table 4.7. These are shown as a percentage of permanent faults. Since equipment faults are almost always permanent faults, the overall percentage of equipment failures is a low percentage of all faults (since most faults are temporary on overhead circuits). On underground circuits, most faults are due to equipment failures.

Distribution transformers are the most common major device, so their failure rate is important. Transformers generally fail at rates of about 0.5% per year. The most common failure mode starts as a breakdown of the turn-to-turn insulation.

Table 4.8 shows equipment failure rates recorded over a 5-year period at PG&E. This data is generic service-time failure rates, which is an estimate to the actual failure rate. Note that this is for California which has very little lightning and mild weather; other areas may have higher equipment failures. The rate of all permanent faults was 0.11 faults/mi/year for rural circuits (0.071 faults/km/year) and 0.16 faults/mi/year for urban circuits (0.102 faults/km/year). The only component where there was a statistical difference between urban and rural at the 90% confidence level was the difference in failure rates of transformers (the sample size for the rest of the numbers was too small to statistically determine a difference).

Another source for equipment failure rate data is the IEEE Gold Book (IEEE Std. 493-1997) (Table 4.9). Note that the Gold Book is for industrial facilities. Application and loading practices may be significantly different than typical utility applications. Still, they provide useful comparisons.

TABLE 4.8

Service-Time Overhead Component Failure Rates for PG & E

Component	Failure Rates per Year	
	Rural	Urban
Transformers	0.0271%	0.0614%
Switches	0.126%	0.0775%
Fuses	0.45%	0.374%
Capacitors	1.05%	0.85%
Reclosers	1.50%	1.44%
Voltage regulators	2.88%	n/a
Conductors	1.22/100 mi	1.98/100 mi

Source: Horton, W. F., Golberg, S., and Volkmann, C. A., "The Failure Rates of Overhead Distribution System Components," IEEE Power Engineering Society Transmission and Distribution Conference, 1991. With permission. ©1991 IEEE.

TABLE 4.9

Overhead Component Failure Rates in the IEEE Gold Book

Component	Failure Rate per Year
Transformers (all)	0.62%
Transformers (300 to 10,000 kVA)	0.59%
Transformers (>10,000 kVA)	1.53%
Switchgear bus (insulated)	0.113%[a]
Switchgear bus (bare)	0.192%[a]

[a] For each circuit breaker and connected switch.

Source: From IEEE Std. 493-1997. Copyright 1998 IEEE. All rights reserved.

4.8 Faults in Equipment

Failures in equipment pose special hazards with important safety ramifications. Transformers deserve extra attention because they are so common. One utility has reported one violent distribution transformer failure for every 270 transformers containing an internal fault, and 20% of re-energizations had internal faults, which is one violent failure every 1350 reenergizations (CEA 149 D 491A, 1997). Cuk (2001) estimated that between 2 and 5% of overhead transformers are re-fused every year (based on analyzing fuse purchases; this number varies with fusing practices). This section discusses failure mechanisms and the consequences of an internal failure.

Transformer insulation degrades over the life of the transformer. Heat drives the degradation of transformer insulation (this is a generality that applies to many other types of insulation as well). Overloading transformers will reduce a transformer's life. Most utilities will overload distribution transformers as it is the best economic way to operate them.

Heat degrades paper insulation at a relatively known rate. ANSI standards give guidelines on loss of life versus temperature. Heating can also cause generation of gas bubbles (Kaufmann and McMillen, 1983). Gas can be created by decomposition of the paper insulation. Prolonged overloading to 175% can cause gas generation. Gas can also be created when heated oil has a pressure drop; the most likely scenario is an overloaded transformer that is cooled quickly (the effect is most significant for overloads above 175%). Rainfall causes the quickest cooling; a loss of load also cools the transformer. The bubbles reduce the dielectric strength of the insulation system. Bubble generation starts when the temperature is near 145°C. Hotspot temperatures exceeding 200°C during overloading can reduce the insulation strength by a factor of two (Kaufmann, 1977). Once the transformer cools off and the bubbles disappear, the insulation recovers most of its initial strength (minus the amount of paper degraded). During overloading, failures can be caused by the power-frequency voltage or a voltage surge (the straw that breaks the camel's back).

Internal faults in equipment such as transformers and capacitors can cause violent damage. Of most concern, explosive failures endanger workers and the public. Figure 4.31 shows a thought-provoking picture of a failure of a recloser. It illustrates how important safety is. Buy quality equipment! Use effective fault protection! Knowing the characteristics of internal failures helps prevent such accidents. We must properly fuse equipment. Fusing should ensure that if equipment does fail internally, it is isolated from the system before it ruptures or ejects any oil.

FIGURE 4.31
Explosion of a recloser that caused a fatality. (From Dalton Sullivan, Pocahontas (AR) Star Herald. With permission.)

During the 1970s, considerable work was done to investigate the failure mechanisms, withstand abilities of transformers, and ways to improve protection (Barkan et al., 1976; Goodman and Zupon, 1976; Mahieu, 1975). The voltage along an arc remains relatively constant regardless of the fault current magnitude when under oil just as it does in air. An arc voltage under oil is roughly 215 to 255 V/in. (85 to 100 V/cm) (Goodman and Zupon, 1976; Mahieu, 1975), which is higher than the voltage gradient of arcs in air. The arc voltage is higher because oil cools the arc, which reduces the ionization. Because the arc voltage is constant, the energy in the transformer is a function of $\int I\,dt$.

During a fault under oil, an arc creates a shock wave in the oil that can cause significant dynamic pressures. Also, considerable gas is generated at the rate of roughly 4.3 to 6.1 $in.^3$/kW (70 to 100 cm^3/kW) (Barkan et al., 1976; Goodman and Zupon, 1976). An arc in oil is hot enough to vaporize oil and ionize the gas. The gasses created by the arc include roughly 65 to 80% hydrogen and 15 to 25% acetylene with ethylene, methane, and higher molecular-weight gasses. The arc also produces considerable amounts of solids and free carbon. The arc generates combustible gasses, but combustion is uncommon because the oxygen level is generally low.

The pressure buildup and failure mode is fairly complicated and depends on the location of the fault. A fault in the windings generally causes less peak pressure on the top of the tank than a free-burning arc in oil even if the energy input is the same. The initial force on the transformer is a large downward force in the oil, but the top usually fails first since that is where the weakest structures are (although the downward force and resulting rebound can cause the transformer to buck violently which may break the supports). The transformer lid is the weakest structural portion, so with excessive pressure, the lid will be the first place to fail. Another common failure is a bushing ejection. For a given arc energy, larger transformers have less pressure buildup because a larger transformer has a larger air space [a 100-kVA transformer has 3.8 times the air volume as a 10-kVA transformer (Barkan et al., 1976)]. Padmounted transformers withstand more pressure buildup than overhead transformers (Benton, 1979). Padmounts have higher volume, and the square shape allows the tank to bulge out, which relieves some of the stress.

Many distribution transformers have pressure release valves. These are not fast enough to appreciably reduce the pressure buildup during high-current faults. The pressure release valves help for low-current faults due to interwinding failures. Most failures of distribution transformers start as interwinding faults, either from turn-to-turn or from layer-to-layer. Turn-to-turn faults on the primary winding draw less than load current, so they will not operate the primary fuse. Turn-to-turn faults on the secondary and layer-to-layer faults on the primary draw higher current that may be high enough to operate the primary fuse. Interwinding faults are low-current events where the pressure builds slowly, so the pressure release valves can effectively release the pressure for primary-winding faults although secondary winding faults may increase pressure faster than the pressure-relief valve can dissipate (Lunsford and Tobin, 1997). As an interwinding fault arcs and causes damage and melts additional insulation, the fault current will increase; usually current jumps sharply to the bolted fault condition (not a slow escalation of current).

Overhead completely self protected transformers (CSPs) and padmounted transformers with under-oil fuses have less withstand capability than conventional transformers. The reason for this is that the under-oil fuse (called a weak-link fuse) provides another arcing location. When the weak-link fuse melts, an arc forms in place of the melted fuse element. This arc is in addition to whatever arc may exist within the tank that caused the fault in the first place. The arc across the fuse location is generally going to be longer than normal arcs that could occur inside a transformer. The length of an under-oil weak-link fuse is 2 to 3 in. (5.1 to 7.6 cm) for a 15-kV class fuse. Higher-voltage transformers have longer fuses — a 35-kV class fuse has a length of about 5 in. (12.7 cm). Also, the voltage gradient along an arc in a fuse tube under oil is greater than a "free" arc in oil [the fuse tube increases the pressure of the arc, which increases the voltage drop (Barkan et al., 1976)].

Transformers with under-oil arresters have a special vulnerability (Henning et al., 1989). The under-oil arrester provides another possible failure mode which can lead to very high energy in the transformer if the arrester fails. If the arrester blocks fail, a relatively long arc results. A 10-kV duty-cycle rated arrester has a total block length of about 4.5 in. (11.4 cm). With such a long arc, the energy in the transformer will be very high. Industry tests and ratings do not directly address this issue. To be conservative, consider using a current-limiting fuse upstream of the transformer if the line-to-ground fault current exceeds 1 or 2 kA if under-oil arresters are used.

Stand-alone arresters are another piece of equipment where failure is a concern. If an arrester fails, a long internal arc may cause the arrester to explode, sending pieces of the housing along with pieces of the metal oxide. The move from porcelain-housed arresters to polymer-housed arresters was motivated primarily by the fact that the polymer-housing is less dangerous if the arrester fails. With a porcelain-housed arrester, the thermal shock from the arc can shatter the housing and forcefully expel the "shrapnel." Polymer-housed arresters are safer because the fault arc splits the polymer housing, which relieves the pressure buildup (although if the arc originates inside the blocks, the pressure can expel bits of the metal oxide). Arresters have caused accidents, and they got a bad name when metal-oxide arresters were first introduced because they occasionally failed upon installation.

Distribution arresters may specify a fault current withstand which is governed by IEEE standards (ANSI/IEEE C62.11-1987). To pass the test, an arrester must withstand an internal failure of the given fault current and all components of the arrester must be confined within the enclosure. The duration of the test is a minimum of 0.1 sec (manufacturers often specify other times as well) which is a typical circuit breaker clearing time when the instantaneous relay element operates. If the available fault current is higher than the rated withstand, then current-limiting fuses should be considered. With polymer-housed arresters, the fault-current withstand is usually sufficient with manufacturers specifying withstand values of 10 to 20 kA for 0.1 sec.

The failure of arresters (especially porcelain-housed arresters) is also a consideration for fusing. If an arrester is downstream of a transformer fuse and the arrester fails, the relatively small transformer fuse will blow. If an arrester is upstream of the transformer fuse then a larger tap fuse or the substation circuit breaker operates, which allows a much longer duration fault current. Arresters have isolators that disconnect the arresters in case of failure. Isolators do *not* clear fault current. After the fuse or circuit breaker operates, the disconnect provides enough separation to allow the circuit to reclose successfully. If the next upstream device is a circuit breaker and an instantaneous element is not used, fault currents could be much longer than the tested 0.1 sec, so consider adding a fuse upstream of the arrester on porcelain-housed units.

References

ABB, *Distribution Transformer Guide*, 1995.

Ackermann, R. H. and Baker, P., "Performance of Distribution Lines in Severe Lightning Areas," International Conference on Lightning and Power Systems, London, U.K., June 1984.

ANSI/IEEE C62.11-1987, *IEEE Standard for Metal-Oxide Surge Arresters for AC Power Circuits*, American National Standards Institute, Institute of Electrical and Electronics Engineers, Inc.

ANSI/IEEE Std. 100-1992, *Standard Dictionary of Electrical and Electronics Terms.*

ASTM F1506-94, *Textile Materials for Wearing Apparel for Use by Electrical Workers Exposed to Momentary Electric Arc and Related Thermal Hazards*, American Society for Testing and Materials, 1994.

Barkan, P., Damsky, B. L., Ettlinger, L. F., and Kotski, E. J., "Overpressure Phenomena in Distribution Transformers with Low Impedance Faults Experiment and Theory," *IEEE Transactions on Power Apparatus and Systems*, vol. PAS-95, no. 1, pp. 37–48, 1976.

Benton, R. E., "Energy Absorption Capabilities of Pad Mounted Distribution Transformers with Internal Faults," IEEE/PES Transmission and Distribution Conference, 1979.

Bernis, R. A. O. and de Minas Gerais, C. E., "CEMIG Addresses Urban Dilemma," *Transmission and Distribution World*, vol. 53, no. 3, pp. 56–61, March 2001.

Burke, J. J. and Lawrence, D. J., "Characteristics of Fault Currents on Distribution Systems," *IEEE Transactions on Power Apparatus and Systems*, vol. PAS-103, no. 1, pp. 1–6, January 1984.

CEA 149 D 491A, *Distribution Transformer Internal Pressure Withstand Test*, Canadian Electrical Association, 1997. As cited by Cuk (2001).

CEA 160 D 597, *Effect of Lightning on the Operating Reliability of Distribution Systems*, Canadian Electrical Association, Montreal, Quebec, 1998.

Cobine, J. D., *Gaseous Conductors*, McGraw-Hill, New York, 1941.

Cuk, N., "Seminar on Detection of Internal Arcing Faults in Distribution Transformers," Presentation at the IEEE Transformers Committee Meeting, Orlando, Florida, October 14–17, 2001.

Drouet, M. G. and Nadeau, F., "Pressure Waves Due to Arcing Faults in a Substation," *IEEE Transactions on Power Apparatus and Systems*, vol. PAS-98, no. 5, pp. 1632–5, 1979.

EPRI 1209-1, *Distribution Fault Current Analysis*, Electric Power Research Institute, Palo Alto, CA, 1983.

EPRI EL-6903, *Study of Fault-Current-Limiting Techniques*, Electric Power Research Institute, Palo Alto, CA, 1990.

EPRI TE-114915, *Mitigation of Animal-Caused Outages for Distribution Lines and Substations*, Electric Power Research Institute, Palo Alto, CA, 1999.

EPRI TR-106294-V2, *An Assessment of Distribution System Power Quality. Volume 2:* Statistical Summary Report, Electric Power Research Institute, Palo Alto, CA, 1996.

EPRI TR-106294-V3, *An Assessment of Distribution System Power Quality. Volume 3:* Library of Distribution System Power Quality Monitoring Case Studies, Electric Power Research Institute, Palo Alto, CA, 1996.

EPRI TR-109178, *Distribution Cost Structure — Methodology and Generic Data*, Electric Power Research Institute, Palo Alto, CA, 1998.

Finch, K., "Understanding Tree Outages," EEI Vegetation Managers Meeting, Palm Springs, CA, May 1, 2001.

Frazier, S. D. and Bonham, C., "Suggested Practices for Reducing Animal-Caused Outages," *IEEE Industry Applications Magazine*, vol. 2, no. 4, pp. 25–31, August 1996.

Girgis, A. A., Fallon, C. M., and Lubkeman, D. L., "A Fault Location Technique for Rural Distribution Feeders," *IEEE Transactions on Industry Applications*, vol. 29, no. 6, pp. 1170–5, November/December 1993.

Goda, Y., Iwata, M., Ikeda, K., and Tanaka, S., "Arc Voltage Characteristics of High Current Fault Arcs in Long Gaps," *IEEE Transactions on Power Delivery*, vol. 15, no. 2, pp. 791–5, April 2000.

Goedde, G. L., Kojovic, L. A., and Knabe, E. S., "Overvoltage Protection for Distribution and Low-Voltage Equipment Experiencing Sustained Overvoltages," Electric Council of New England (ECNE) Fall Engineering Conference, Killington, VT, 2002.

Goodfellow, J., "Understanding the Way Trees Cause Outages," 2000. http://www.eci-consulting.com/images/wnew/trees.pdf.

Goodman, E. A. and Zupon, L., "Static Pressures Developed in Distribution Transformers Due to Internal Arcing Under Oil," *IEEE Transactions on Power Apparatus and Systems*, vol. PAS-95, no. 5, pp. 1689–98, 1976.

Hart, B., "HV Overhead Line — the Scandinavian Experience," *Power Engineering Journal*, pp. 119–23, June 1994.

Hendrix, "Reliability of Overhead Distribution Circuits," Hendrix Wire & Cable, Inc., August 1998.

Henning, W. R., Hernandez, A. D., and Lien, W. W., "Fault Current Capability of Distribution Transformers with Under-Oil Arresters," *IEEE Transactions on Power Delivery*, vol. 4, no. 1, pp. 405–12, January 1989.

Horton, W. F., Golberg, S., and Volkmann, C. A., "The Failure Rates of Overhead Distribution System Components," IEEE Power Engineering Society Transmission and Distribution Conference, 1991.

IEEE C2-1997, *National Electrical Safety Code.*

IEEE Std. 493-1997, *IEEE Recommended Practice for the Design of Reliable Industrial and Commercial Power Systems (Gold Book).*

IEEE Std. 1024-1988, *IEEE Recommended Practice for Specifying Distribution Composite Insulators (Suspension Type).*

IEEE Tutorial Course 90EH0310-3-PWR, "Detection of Downed Conductors on Utility Distribution Systems," 1990.

IEEE Working Group on Distribution Protection, "Distribution Line Protection Practices Industry Survey Results," *IEEE Transactions on Power Delivery,* vol. 10, no. 1, pp. 176–86, January 1995.

Johnston, L., Tweed, N. B., Ward, D. J., and Burke, J. J., "An Analysis of Vepco's 34.5 kV Distribution Feeder Faults as Related to Through Fault Failures of Substation Transformers," *IEEE Transactions on Power Apparatus and Systems,* vol. PAS-97, no. 5, pp. 1876–84, 1978.

Kaufmann, G. H., "Impulse Testing of Distribution Transformers Under Load," *IEEE Transactions on Power Apparatus and Systems,* vol. PAS-96, no. 5, pp. 1583–95, September/October 1977.

Kaufmann, G. H. and McMillen, C. J., "Gas Bubble Studies and Impulse Tests on Distribution Transformers During Loading Above Nameplate Rating," *IEEE Transactions on Power Apparatus and Systems,* vol. PAS-102, no. 8, pp. 2531–42, August 1983.

Kersting, W. H. and Phillips, W. H., "Modeling and Analysis of Unsymmetrical Transformer Banks Serving Unbalanced Loads," *IEEE Transactions on Industry Applications,* vol. 32, no. 3, pp. 720–5, May-June 1996.

Kuntz, P. A., "Optimal Reliability Centered Vegetation Maintenance Scheduling in Electric Power Distribution Systems," Ph.D. thesis, University of Washington, 1999. As cited by Brown, R. E., *Electric Power Distribution Reliability,* Marcel Dekker, 2002.

Lampley, G. C., "Fault Detection and Location on Electrical Distribution System Case Study," IEEE Rural Electric Power Conference, 2002.

Lasseter, J. A., "Burndown Tests on Bare Conductor," *Electric Light & Power,* pp. 94–100, December 1956.

Lasseter, J. A., "Point Way to Reduce Lightning Outages," *Electrical World,* pp. 93–5, October 1965.

Laverick, D. G., "Performance of the Manweb Rural 11-kV Network During Lightning Activity," International Conference on Lightning and Power Systems, London, June 1984.

Littler, G. E., "High Speed Reclosing and Protection Techniques Applied to Wood Pole Lines," *Elec. Eng. Trans., Inst. Eng. Aust.,* vol. EE1, pp. 107–16, June 1965.

Lunsford, J. M. and Tobin, T. J., "Detection of and Protection for Internal Low-Current Winding Faults in Overhead Distribution Transformers," *IEEE Transactions on Power Delivery,* vol. 12, no. 3, pp. 1241–9, July 1997.

Mahieu, W. R., "Prevention of High-fault Rupture of Pole-Type Distribution Transformers," *IEEE Transactions on Power Apparatus and Systems,* vol. PAS-94, no. 5, pp. 1698–707, 1975.

Mazurek, B., Maczka, T., and Gmitrzak, A., "Effect of Arc Current on Properties of Composite Insulators for Overhead Lines," *Proceedings of the 6th International Conference on Properties and Applications of Dielectric Materials,* June 2000.

Mo, Y. C. and Taylor, L. S., "A Novel Approach for Distribution Fault Analysis," *IEEE Transactions on Power Delivery,* vol. 8, no. 4, pp. 1882–9, October 1993.

Mo, Y. C. and Taylor, L. S., "Analysis and Prevention of Animal-Caused Faults in Power Distribution Systems," *IEEE Transactions on Power Delivery*, vol. 10, no. 2, pp. 995–1001, April 1995.

NFPA 70E-2000, *Standard for Electrical Safety Requirements for Employee Workplaces*, National Fire Protection Association.

Parrish, D. E., "Lightning-Caused Distribution Circuit Breaker Operations," *IEEE Transactions on Power Delivery*, vol. 6, no. 4, pp. 1395–401, October 1991.

Peele, S., "Faults Per Feeder Mile," PQA Conference 2001.

PTI, "Distribution Transformer Application Course Notes," Power Technologies, Inc., Schenectady, NY, 1999.

Rizk, F. A. and Nguyen, D. H., "AC Source-Insulator Interaction in HV Pollution Tests," *IEEE Transactions on Power Apparatus and Systems*, vol. PAS-103, no. 4, pp. 723–32, April 1984.

Santoso, S., Dugan, R. C., Lamoree, J., and Sundaram, A., "Distance Estimation Technique for Single Line-to-Ground Faults in a Radial Distribution System," IEEE Power Engineering Society Winter Meeting, 2000.

Schweitzer, E. O., "A Review of Impedance-Based Fault Locating Experience," Fourteenth annual Iowa-Nebraska system protection seminar, Omaha, NE, October 16, 1990. Downloaded from www.schweitzer.com.

Short, T. A., "Distribution Fault Rate Survey," EPRI PEAC, 2001.

Simpson, P., "EUA's Dual Approach Reduces Tree-Caused Outages," *Transmission & Distribution World*, pp. 22–8, August 1997.

Slepian, J., "Extinction of a Long AC Arc," *AIEE Transactions*, vol. 49, pp. 421–30, April 1930.

Smith, D. R., "System Considerations — Impedance and Fault Current Calculations," *IEEE Tutorial Course on Application and Coordination of Reclosers, Sectionalizers, and Fuses, 1980.* Publication 80 EHO157-8-PWR.

St. Pierre, C., *A Practical Guide to Short-Circuit Calculations*, Electric Power Consultants, Schenectady, NY, 2001.

Strom, A. P., "Long 60-Cycle Arcs in Air," *AIEE Transactions*, vol. 65, pp. 113–8, March 1946.

Taylor, L., "Elimination of Faults on Overhead Distribution Systems," IEEE/PES Winter Power Meeting, 1995. Presentation to the Working Group on System Design, 1995 IEEE Winter Power Meeting, New York, NY.

Williams, C., "Tree Fault Testing on a 12 kV Distribution System," IEEE/PES Winter Power Meeting presentation to the Distribution Subcommittee, 1999.

You begin to think that trees possibly grow faster toward your power line just out of spite.

Powerlineman law #47, By CD Thayer and other Power Linemen
http://www.cdthayer.com/lineman.htm

5

Reliability and Power Quality Improvement Programs

Tom Short, EPRI Solutions, Inc.
Lee Taylor, Duke Power

The three most significant power quality concerns for most utility customers are: sustained interruptions, momentary interruptions, and voltage sags. These three problems are caused by faults on the utility power system, with most of them on the distribution system. The two main strategies to improve power quality to customers are:

- *Minimize the effect of faults on customers* — Sectionalize and restore circuits faster (automation), use more protective devices (more fuses, more reclosers), reclose faster, improve coordination, etc.
- *Eliminate faults* — Better tree maintenance, animal protection, equipment replacement programs, better arrester application, construction work audits to ensure quality, line inspections, etc.

The most effective reliability and power quality improvement programs apply these strategies to target improved circuit performance for maximum customer benefit.

5.1 Improvements in Protection Practices

Improvements in the application of overcurrent protective devices can improve utility reliability and power quality by minimizing the effect of faults on customers. Options include using more protective devices that can sectionalize and restore circuits faster (automation), using more protective devices (more fuses, more reclosers), reclosing faster, improving coordination, and so on. An advantage of optimizing protection as a way of improving reliability is that many options are low cost — utilities can improve performance by changing relay or recloser settings or applying low-cost fuses more appropriately.

5.1.1 Fusing

Fuses — the more we have, the more we isolate faults to smaller chunks of circuitry, and the fewer customers we interrupt. Taps are almost universally fused, primarily for reliability. Fuses make cheap fault finders. We want to have a high percentage of a circuit's exposure on fused taps so that, when permanent faults occur on those sections, only a small number of customers are interrupted. A number of overlooked scenarios can be improved with better application of fuses: (1) unfused taps on the mainline; (2) transformers without a local fuse; and (3) arresters upstream of fuses.

Many utilities have more taps that are unfused than they realize. If the unfused taps use a smaller conductor size, faults are more likely to burn down such conductors because a circuit breaker will take longer to clear a fault than would a fuse. Because such taps may be on side streets, during patrols, crews may forget to inspect them for damage, increasing the interruption time. Therefore, one way to improve reliability is to make sure that *all* taps are fused. Consider a common example in which an unfused tap crosses the street to feed a riser pole. To protect the small, unfused tap, move the fuses from the riser pole to the pole where the circuit taps off the mainline.

Having a local external fuse to protect a transformer helps improve reliability. If the transformer is a completely self-protected transformer (CSP, see Figure 5.1) with an internal fuse, then an animal across a bushing or

FIGURE 5.1
Unfused CSP on a mainline.

other bushing failure will force the tap fuse or upstream circuit breaker or recloser to operate, leaving many more customers interrupted, with much more area for crews to patrol. CSPs on the mainline are especially problematic for reliability indices. A local fuse also helps crews more accurately find and identify the source of the problem. The most common fusing equipment for this application is an expulsion fuse in a cutout, but current limiting fuses are also an option. Because CSPs have lower structural withstand capabilities than conventional transformers, current limiting fuses are appropriate in many locations to provide protection against violent transformer failure (see Short, 2004, for more information).

Arresters should be placed downstream of fuses if possible. Then, an arrester failure or an animal across the arrester will blow the fuse rather than forcing a mainline protective device operation (interrupting many more customers than if a fuse had operated). If the arrester isolator fails to operate (which can happen), the failure may be extremely hard for the crews to find. A fuse helps to localize the failure.

Arresters upstream of fuses can cause further problems. If the arrester fails and the isolator operates, crews may reclose the circuit successfully if they do not find the failed arrester. This leaves the equipment unprotected. Worse yet, the failed arrester body may start to track across the bracket. Eventually, the arrester bracket will flash over, causing a permanent fault that is difficult to find. With arresters upstream of a fuse, arrester lead lengths will be longer, and the equipment is not protected as well against lightning surges. This can lead to more lightning-caused equipment failures.

5.1.2 Fuse Saving vs. Fuse Blowing

Reviewing circuit breaker and fuse coordination is another approach to improving reliability. Circuits usually employ fuse saving or fuse blowing. With fuse saving, the instantaneous relay element on a breaker or the fast curve on a recloser clears faults before downstream lateral fuses operate. Because most faults on overhead distribution circuits are temporary, fuse saving is used to improve reliability by reducing tap fuse operations. With fuse blowing, the fuse coordinates to blow always before the upstream breaker operates, greatly reducing the number of momentary interruptions on the circuit. About one third of utilities use fuse saving, one third use fuse blowing, and one third use combinations of strategies with the trend towards moving to fuse blowing (Short, 1999).

To estimate the change in the number of momentary interruptions with fuse saving, simply use the ratio of the length of the mains to the total length of the circuit including all laterals. For example, if a circuit has 5 miles of mains and 10 miles of laterals, the number of momentaries after switching to fuse blowing would be one third of the number of momentaries with fuse saving ($5/(5 + 10) = 1/3$). This assumes that the mains and laterals have the

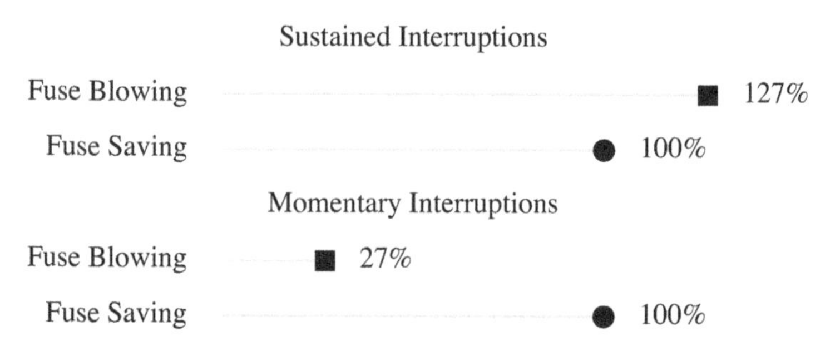

FIGURE 5.2
Comparison of fuse saving and fuse blowing on a hypothetical circuit. Mains: 10 miles (16.1 km), fault rate=0.5/mile/year (0.8/km/year), 75% temporary. Taps: 10 miles (16.1 km) total, 20 laterals, fault rate=2/mile/year (3.2/km/year), 75% temporary. It also assumes that fuse saving is 100% successful.

same fault rate. If the fault rate on laterals is higher (which it often is because of less tree trimming, etc.), the number of momentaries is even less.

Note how dramatically momentaries can be reduced by using fuse blowing. No other method can so easily eliminate 30 to 70% of momentaries. The effect on reliability of going to a fuse-blowing scheme is more difficult to estimate. Fuse blowing increases the number of fuse operations by 40 to 500% (Dugan et al., 1996; Short, 1999; Short and Ammon, 1997). This will increase the average frequency of sustained interruptions by 10 to 60%, though many variables can change the ratios. One example is given in Figure 5.2. Note that the effect on sustained interruptions is not equally distributed. Customers on the mains see no difference in the number of permanent interruptions (but will notice the difference in momentary interruptions). Customers on long laterals may have many more sustained interruptions with a fuse-blowing scheme.

One of the main reasons that utilities have decided not to use fuse saving is that it is difficult to make it work. Fuses clear quickly relative to circuit breakers, so when fault currents are high, the fuse blows before the breaker trips. This results in a fuse operation and a momentary interruption for all customers on the circuit. K links, the most common lateral fuses, are fast fuses. Most distribution circuit breakers take five cycles to clear. For fuse saving to work, the breaker must open before the fuse blows, so the fuse needs to survive for the time that it takes the instantaneous relay to operate (about one cycle) plus the five cycles for the breaker. For example, with a five-cycle breaker and a 100-K fuse, fuse saving only coordinates for fault currents below 1354 A (Short, 2004). Smaller fuses have lower current limits.

Fuse blowing also has drawbacks — faults on the mains can last a long time. With fuse saving, main-line faults normally clear in five to seven cycles (0.1 sec) on the first shot with the instantaneous element. With fuse blowing, this same fault may last for 0.5 to 1 sec. Much more damage at the fault location occurs during this extra time. Problems include:

- *Conductor burndowns* — At the fault, the heat from the fault current arc burns the conductor enough to break it, dropping it to the ground.
- *Damage of inline equipment* — The most common problem has been with inline hot-line clamps. If the connection is not good, the high-current fault arc across the contact can burn the connection apart.
- *Station transformers* — Longer faults subject substation transformers to extra duty.
- *Evolving faults* — Ground faults are more likely to become two- or three-phase faults. Faults on underbuilt distribution are more likely to cause faults on the transmission circuit above due to rising arc gases.

In order for a fuse-saving scheme to work, it is necessary for the substation protective device to open before fuses operate. This can be achieved in several ways:

- *Slow down the fuse* — Use big, slow fuses (such as a 140 or 200 T) near the substation to ensure proper coordination.
- *Faster breakers or reclosers* —If three-cycle circuit breakers are used instead of the normal five-cycle breakers, fuse-saving coordination is more likely. Some reclosers are even faster than three-cycle breakers.
- *Limit fault currents*:
  - Open station bus ties: an open bus tie will reduce the fault current on each feeder and make fuse saving easier. This is the normal operating mode for most utilities.
  - Use a transformer neutral reactor: a neutral reactor reduces the fault current for single-phase faults (all faults on single-phase taps).
  - Use line reactors: this reduces the fault current for all types of faults. This has been an uncommon practice. An added advantage is that reactors reduce the impact of voltage sags for faults on adjacent feeders.
  - Specify higher impedance transformers.

With fuse saving, consider other strategies to limit the impact of momentary interruptions:

- *More downstream reclosers* — Extra downstream devices will reduce the number of momentaries for customers near the substation. It is important to coordinate reclosers with the upstream device (including sequence coordination).

- *Single-phase reclosers*
- *Immediate reclose*
- *Switch to a fuse blowing scheme on poor feeders* — For a feeder with many momentaries, disable the instantaneous relay for a time period. Identify poorly performing parts of the circuit during this time. The blown branch fuses provide a convenient fault location method. Once the poorly performing sections are identified and improved, switch the circuit back to fuse saving.

Several strategies can optimize a fuse-blowing scheme:

- *Fast fuses (or current-limiting fuses)* — If smaller or faster fuses are used, faults clear faster, so voltage sag durations are shorter. Current-limiting fuses also limit the magnitude and duration of the sag. Be careful not to fuse too small because fuses will operate unnecessarily due to loading, inrush, and cold load pickup. Note that if smaller fuses are used, it is difficult to switch back to a fuse saving scheme.
- *Covered wire or small wire* — Watch burndowns on circuits with covered wire or small wire protected by the station circuit breaker or recloser. If either of these cases exists, use a modified fuse-blowing scheme with a time-delayed instantaneous element (see Short, 2004).
- *Use single-phase reclosers on longer laterals* — A good way of maintaining some of the reliability of a fuse-saving scheme is to use single-phase reclosers instead of fuses on longer taps. Then, temporary faults on these laterals do not cause permanent interruptions to those customers.
- *More fuses* — Add more second- and third-level fuses to segment the circuit more.
- *Track lateral operations* — Temporary faults on fused laterals cause sustained interruptions. In order to minimize the impacts on lateral customers, track interruptions by lateral. Identify poorly performing laterals, patrol poor sections, and then add tree trimming, animal guards, etc.

Neither a fuse-saving nor a fuse-blowing protection scheme is the best choice for all applications. One is better in some applications than others. The best choice depends on many factors, including fusing practices, wire type used, the mix and location of customers on a circuit, and the utility philosophy. It is helpful to review the choice made (even on a circuit-by-circuit basis) because many situations would be better served by a different choice. For a more detailed review of options, see Short (1999, 2004).

5.1.3 Reclosing Practices

Optimizing reclosing strategies is another way to minimize the impact of faults to customers. Choosing the number of reclosing attempts is a balancing act. Each reclosing attempt has some chance that the circuit will hold, but the probability diminishes quickly with each successive attempt. Excessive reclosing increases stress on the system:

- *Additional damage at the fault location* — With each reclose into a fault, arcing does additional damage at the fault location. Faults in equipment do more damage; cable faults are harder to splice, wire burndowns are more likely, and oil-filled equipment is more likely to rupture. Arcs can start fires. Faults (and the damage that the arcs cause) can propagate from one phase to other phases.
- *Voltage sags* — With each reclose into a fault, customers on adjacent circuits are hit with another voltage sag. It can be argued that the magnitude and duration of the sag should be about the same; thus, depending on the type of device, if it survived the first sag, it will probably ride through subsequent sags of the same severity. If additional phases become involved in the fault, the voltage sag is more severe.
- *Through-fault damage to transformers* — Each fault subjects transformers to mechanical and thermal stresses. Virginia Power changed its reclosing practices because of excessive transformer failures on its 34.5-kV station transformers due to through faults (Johnston et al., 1978).
- *Through-fault damage to other equipment* — Cables, wires, and, especially, connectors suffer the thermal and mechanical stresses of the fault.
- *Interrupt ratings of breakers* — Circuit breakers must be derated if the reclose cycle involves more than one reclose attempt within 15 sec. This may be a consideration if fault currents are high and breakers are near their ratings. Reclosers do not have to be derated for a complete four-sequence operation. Extra reclose attempts increase the number of operations, which means more frequent breaker and recloser maintenance.
- *Ratcheting of overcurrent relays* — An induction relay disc turns in response to fault current. After the fault is over, it takes time for the disc to spin back to the neutral position. If this reset is not completed, and another fault occurs, the disc starts spinning from its existing condition, making the relay operate faster than it should. The most common problem area is miscoordination of a substation feeder relay with a downstream feeder recloser. If a fault occurs downstream of the recloser, the induction relay will spin due to the current (but not operate if it is properly coordinated). Multiple recloses by

the recloser could ratchet the station relay enough to falsely trip the relay. The normal solution is to take the ratcheting into account when coordinating the relay and recloser; however, in some cases modification of the reclosing cycle of the recloser is an option. Another option is to use digital relays, which do not ratchet in this manner.

Given these concerns, the trend has been to decrease the number of reclose attempts. Try to balance the loss in reliability against the problems caused by extra reclose attempts. A major question is how often the extra reclose attempts are successful. Table 5.1 shows the success rate for one utility in a high-lightning area. The key point is that it is relatively uncommon for the third or fourth reclose attempt to be successful.

Some engineers and field personnel believe that the purpose of the extra reclose attempts is to *burn* the fault clear. This is dangerous. Faults regularly burn clear on low-voltage systems (≤480 V), but rarely at distribution primary voltages. Faults can burn clear on primary systems — the most common example is that tree branches or animals can be burned loose. The problem with this concept is that, just as easily, the fault burns the primary conductor, which falls to the ground causing a high-impedance fault. Fires and equipment damage are also more likely with the "burn clear" philosophy.

To reduce the impacts of subsequent reclose attempts, one could switch back to an instantaneous operation after the first time-overcurrent relay operation. If the fault does not clear then, it means the fault is not downstream of a fuse (or a recloser). The reason to use a time-overcurrent relay is to coordinate with the fuse. Because the fuse is out of the picture, why not use a faster trip for subsequent reclose attempts? Although this is not commonly done, it could be easily implemented with digital relays. The setting of the "subsequent reclose" instantaneous relay element should be different from that of the first-shot instantaneous. Set the pickup at the pickup of the time-overcurrent relay. Because of inrush on subsequent attempts, a fast time-overcurrent curve or an instantaneous element with a short delay (approximately five cycles) may be used.

As an example, if a utility uses a 0–15–30–90-sec reclosing cycle, the system is subjected to five faults if the system goes through its complete cycle. With the instantaneous operation enabled on the first attempt and disabled on subsequent attempts, a very high total duration of the fault current is present.

TABLE 5.1

Reclose Success Rates For a Utility in a High Lightning Area

Reclosure	Success rate	Cumulative success
1st shot (immediate)	83.25%	83.25%
2nd shot (15 to 45 seconds)	10.05%	93.30%
3rd shot (120 seconds)	1.42%	94.72%
Locked out	5.28%	

Data source: Westinghouse [1982]

For a CO-11 ground relay with a time-dial of 3, a 2-kA fault clears in roughly 1 sec. For the reclosing cycle to lock out, the system has a total fault time of 4.1 sec (one 0.1-sec fault followed by four 1-sec faults). If the instantaneous operation is enabled for reclose attempts two through four, the total fault duration is 1.4 sec (one 0.1-sec fault followed by a 1-sec fault and three 0.1-sec faults). This greatly reduces the damage done by certain faults.

One scenario that can disrupt this approach is with downstream single-phase hydraulic reclosers on three-phase lines. The problem occurs when a permanent fault spreads from one phase to the other, leaving the line reclosers slightly out of sequence. The substation time-overcurrent relay is finished, but one line recloser is still not quite through its C curve. After the second reclose of the substation breaker, the downstream recloser is still not quite open, and the instantaneous operation takes the entire circuit out for a fault beyond the line reclosers. The evidence can be seen on the sequence-of-events details because the fault is obviously coming from beyond the line recloser. It might appear that the line recloser is malfunctioning (i.e., too slow), but it is actually caused by the evolving fault. Therefore, Duke Power enables the instantaneous operation after the time-overcurrent element only if no single-phase hydraulic reclosers are downstream (on three-phase lines). This is not a concern with fuses or electronic three-phase reclosers.

Using an "immediate" reclose attempt on the first shot is a way to reduce the impact to customers. An immediate reclose means having no intentional time delay (or a very short time delay) on the first reclose attempt on circuit breakers and reclosers. Many residential devices, such as digital clocks, DVD players, and microwaves, can ride through a 0.5-sec interruption but not a 5-sec interruption (see Chapter 2), so a fast reclose helps reduce residential complaints.

The most common power quality recorder in the world is the digital clock. Many complaints are due to "blinking clocks." Using an immediate reclose reduces complaints. Florida Power Corp. has reported that a reclosing time of 18 to 20 cycles nearly eliminates complaints (Dugan et al., 1996). Another utility that has successfully used the immediate reclose is Long Island Lighting Company (now LIPA) (Short and Ammon, 1997). According to an IEEE survey, a time of less than 1 sec to first reclose is the most common practice, although the fast reclose practice tends to decline with increasing voltage (IEEE Working Group on Distribution Protection, 1995).

Sometimes a delayed reclose is necessary if there is not enough time to clear the fault. Whether the arc strikes again is a function of voltage and structure spacings. The deionization time increases only moderately with voltage. Even for a 34.5-kV system, the deionization time is 11.5 cycles (Westinghouse Electric Corporation, 1982). The reclose time for distribution circuit breakers and reclosers varies by design. A typical time is 0.4 to 0.6 sec for an immediate reclose (meaning no intentional delay). The fastest devices (newer vacuum or SF_6 devices) may reclose in as little as 11 cycles. This may prove to be too fast for some applications, so consider adding a small delay of 0.1 to 0.4 sec (especially at 25 or 35 kV).

On distribution circuits, other things affect the time to clear a fault besides the deionization of the arc stream. If a temporary fault is caused by a tree limb or animal, time may be needed for the "debris" to fall off the conductors or insulators. Because of this, with an immediate reclose, at least two reclose attempts should be made before lockout. For example, use a 0–15–30-sec cycle (three reclose attempts) or, if two reclose attempts are to be made, use a 0–30- or 0–45-sec cycle (use a long delay before the last reclose attempt).

5.1.4 Single-Phase Protective Devices

Single-phase protective devices on three-phase circuits can help on residential circuits by only interrupting one third of the customers for single-phase faults. Many distribution protective devices are single phase or are available in single-phase versions, including reclosers, fuses, and sectionalizers. Single-phase protective devices are used widely on distribution systems. Single-phase taps are almost universally fused. On long single-phase taps, single-phase reclosers are sometimes used. Most utilities also use fuses for three-phase taps, and some also fuse portions of three-phase mainlines. The utilities that do not fuse three-phase taps most often cite the problem of single-phasing motors of three-phase customers. Some utilities use single-phase reclosers that protect three-phase circuits, even in the substation.

Single-phase protective devices on single-phase laterals are widely used, and the benefits are universally accepted. The fuse provides an inexpensive way of isolating faulted circuit sections and also aids in finding the fault.

Using single-phase interrupters helps improve reliability on three-phase circuits: only one phase is interrupted for line-to-ground faults. The effect on individual customers can be easily estimated using the number of phases that are faulted on average, as shown in Table 5.2. Overall, using single-phase protective devices cuts the average number of interruptions in half. This assumes that all customers are single phase and that the customers are evenly split between phases.

Service to three-phase customers downstream of single-phase interrupters generally improves too. Three-phase customers have many single-phase

TABLE 5.2

Effect on Interruptions When Using Single-Phase Protective Devices on Three-Phase Circuits

Fault type	Percent of faults	Portion affected	Weighted effect
Single phase	70%	33%	23%
Two phase	20%	67%	13%
Three phase	10%	100%	10%
		TOTAL	47%

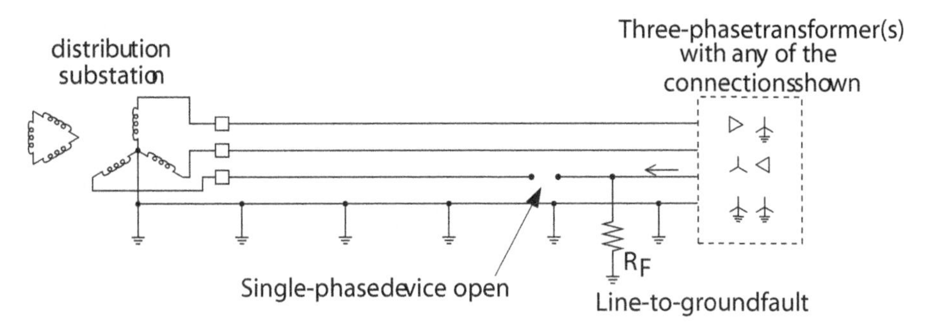

FIGURE 5.3
Single-phase device operation with backfeed.

loads, and the loads on the unfaulted phases are unaffected by the fault. Three-phase devices may also ride through an event caused by a single-phase fault (although motors may heat up because of the voltage unbalance, as discussed in the next section). Single-phase protective devices do have some drawbacks. The main concerns are ferroresonance, backfeeds, and single phasing of motors.

5.1.4.1 Ferroresonance

Ferroresonance usually occurs during manual switching of single-pole switching devices (where the load is usually an unloaded transformer). It is less common for ferroresonance to occur downstream of a single-phase protective device operating due to a fault. The reason for this is that if there is a fault on the opened phase, the fault prevents an overvoltage on the opened phase. Also, any load on the opened section helps prevent ferroresonant overvoltages. Because ferroresonance will be uncommon with single-phase protective devices, it is usually not a major factor in protective device selection. Still, caution is warranted on small three-phase transformers that may be switched unloaded (especially at 24.94 or 34.5 kV).

5.1.4.2 Backfeeds

With single-phase protective devices, backfeeds can create hazards (Figure 5.3). During a line-to-ground fault in which a single-phase device opens, backfeed through a three-phase load can cause voltage on the load side of an opened protective device. Backfeeds can happen with most types of three-phase distribution transformer connections (even with a grounded-wye–grounded-wye connection). Based on a more complete review (Short, 2004), the important points found are:

- The backfeed voltage is enough to be a safety hazard to workers or the public (for example, in a wire-down situation).

- The available backfeed is a stiff enough source to maintain an arc of significant length. The arc can continue causing damage at the fault location during a backfeed condition. It may also be a low-level sparking and sputtering fault.

Based on these points, single phasing can cause problems from backfeeding. Whether this constrains use of single-phase protective devices is debatable. Most utilities do use single-phase protective devices, usually with fuses, on three-phase circuits.

5.1.4.3 Single-Phasing Impacts on Motors

Under single phasing, motors can overheat and fail. Motors have relatively low impedance to negative-sequence voltage; therefore, a small negative-sequence component of the voltage produces a relatively large negative-sequence current. Consequently the effect magnifies; a small negative-sequence voltage appears as a significantly larger percentage of unbalanced current than the percentage of unbalanced voltage.

Loss of one or two phases is a large unbalance. With one phase open, a motor will draw a negative-sequence current equal to the positive-sequence current (and equal to the positive-sequence current prior to the open-phase condition). Negative-sequence current in the stator causes significantly more heating in the rotor of the motor; the field from this current rotates in the opposite direction and forces the rotor current to flow in a smaller area on the rotor. Zocholl (2003) gives the following approximation for the heating from an open-phase condition (Blackburn, 1987 provides a similar expression):

$$I_l^2 + 51_2^2 = 61_1^2$$

where

I_1 = positive-sequence motor current
I_2 = negative-sequence motor current

The negative-sequence current causes significantly more heating than the positive-sequence current; in total, the heating effect of an open phase is equivalent to having five to six times the heating prior to the open phase. For a fully loaded motor, this is equivalent to locked-rotor heating. Motor manufacturers typically provide hot and cold locked-rotor times of between 10 and 20 sec, which is on the order of thermal limits for an open-phase condition on a fully loaded motor. As a reference point, Zocholl (2003) recommends a 4-sec protective relay time-delay setting for a negative-sequence overcurrent relay on a motor to protect against open-phase conditions.

Most utility service agreements with customers state that it is the customer's responsibility to protect his or her equipment against single phasing. The best way to protect motors is with a phase-loss relay. Nevertheless, some utilities take measures to reduce the possibility of single phasing customer's motors, and one way to do that is to limit the use of single-phase protective devices. Other utilities are more aggressive in their use of single-phase protective equipment and leave it up to customers to protect their equipment.

5.1.4.4 Single-Phase Trip, Three-Phase Lockout

Many single-phase reclosers and recloser controls come with a controller option for a single-phase trip and three-phase lockout. Three-phase reclosers that can operate each phase independently are also available. For single-phase faults, only the faulted phase opens. For temporary faults, the recloser successfully clears the fault and closes back in, so the interruption on the faulted phase is only momentary. If the fault is still present after the final reclose attempt (a permanent fault), the recloser trips all three phases and will not attempt additional reclosing operations.

Problems of single-phasing motors, backfeeds, and ferroresonance disappear. Single-phasing motors and ferroresonance cause heating, and heating usually takes many minutes for damage to occur. Short-duration single phasing occurring during a typical reclose cycle does not cause enough heating damage. If the fault is permanent, all three phases trip and lock out, so there is no long-term single phasing. A three-phase lockout also reduces the chance of backfeed to a downed wire for a prolonged period.

5.1.5 Improving Coordination

Other protection-related strategies are also valuable. Automated switches or recloser loops can automatically sectionalize a circuit after a permanent fault, greatly decreasing the restoration time. Reviewing and improving protective device coordination can reduce the number of customers affected by a fault.

Some protection scenarios have long been difficult to coordinate. When miscoordinations occur, more customers are interrupted than necessary. Digital relays and recloser controls offer more options to avoid miscoordinations. Digital relays avoid relay overtravel and ratcheting that can make coordination more difficult on electromechanical relays.

One common miscoordination is between the fast curves (or instantaneous elements) of breakers and reclosers (or between two reclosers). *Sequence coordination* allows easier coordination between devices. Sequence coordination is available on electronic reclosers and also on digital relays controlling circuit breakers. With sequence coordination, the station device detects and counts faults — but does not open — for a fault cleared by a downstream protector on the fast trip. If the fault current occurs again (usually because the fault is permanent), the station device switches to the time-overcurrent element because it counted the first as an operation. Using this form of coordination eliminates the momentary interruption for the entire feeder for permanent faults downstream of a feeder recloser.

On a relay or recloser that has sequence coordination, if the device senses current above some minimum trip setting and the current does not last long enough to trip based on the device's fast curve, the device advances its control-sequence counter as if the unit had operated on its fast curve. Thus, when the downstream device moves to its delayed curve, the upstream device with sequence coordination also is operating on its delayed curve. This allows the upstream device still to provide backup protection in case the downstream device fails. With sequence coordination, for the fast curves, the response curve of the upstream device must still be slower than the clearing curve of the downstream device. Even if a device capable of sequence coordination at the substation is not available, a similar result can be achieved by making sure that the zone of protection of the upstream device has only a small overlap with the downstream device. Urban circuits often have the entire circuit within the range of the station instantaneous relay, so one can reduce momentary interruptions by changing the setting of the instantaneous relay element at the station so that it would not see way past the downstream line reclosers.

Utilities can also use more advanced features of digital relays to improve reliability. Fault recordings from digital relays can help reveal problems, including miscoordinations and malfunctioning devices. Fault location algorithms can help crews find faults more quickly; this will speed restoration, especially for mainline lockouts on long circuits.

5.1.6 Locating Sectionalizing Equipment

To locate and prioritize sectionalizing projects, we can estimate the number of customer interruptions avoided by the addition of the recloser, fuse, or other sectionalizing device. The main inputs for this calculation are fault rates and circuit lengths. Consider the common but simple example of applying a recloser at the midpoint of a circuit mainline in Figure 5.4a. The recloser protects the customers on the upstream side of the recloser from faults on the mainline downstream of the recloser. To estimate the number of customer interruptions saved, multiply the number of customers affected (all of those customers upstream of the recloser) by the fault rate times the mainline exposure downstream of the recloser.

If one were finding locations for reclosers, the number of customers saved by each possible recloser location allows one to prioritize possible recloser locations. When comparing different equipment options, the cost effectiveness of each application can be determined by using the ratio of the cost of the installation of the sectionalizing device to the number of customer interruptions saved. Then, select projects with the lowest cost per customer interruption saved.

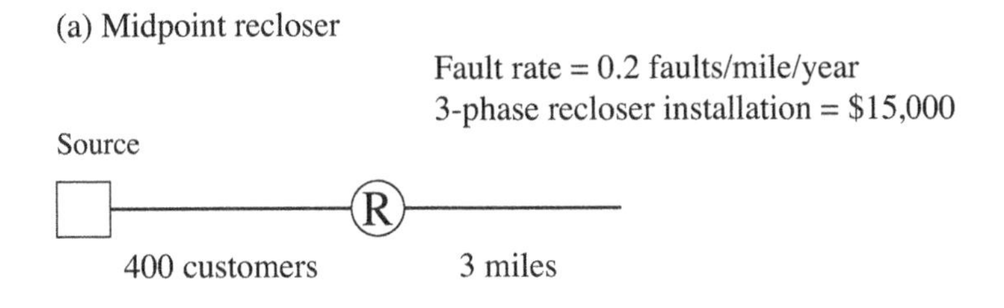

Customer-interruptions saved
= 400 customers (3 miles) (0.2 faults/mile)
= 240 customer interruptions
Cost per customer interruption saved

$$= \frac{\$15{,}000}{240 \text{ cust. int. saved}} = \$62.5 \text{ / cust. int. saved}$$

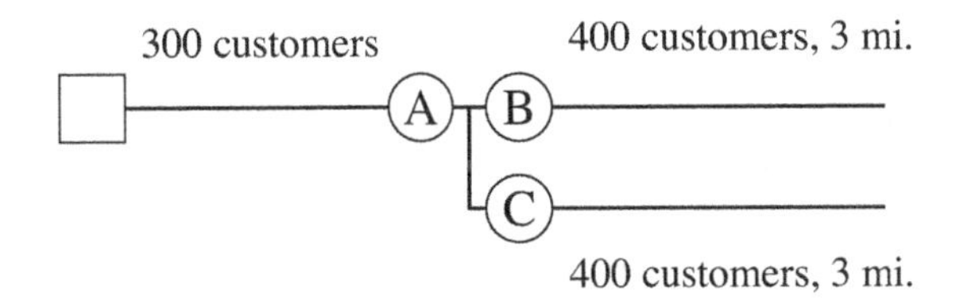

Option A: 300 cust (6 mi) (0.2 faults/mi) = 360 cust. int. saved
Option B or C: 700 cust (3 mi) (0.2 faults/mi) = 420 cust. int. saved
Option C with A already there: 400 cust (3 mi) (0.2 faults/mi)
= 240 cust. int. saved
Option C with B already there: 700 cust (3 mi) (0.2 faults/mi)
= 420 cust. int. saved
Costs per customer interruption saved:
Option A: $15,000 / 360 = $ 41.7 / cust. int. saved
Option B or C: $15,000 / 420 = $ 35.7 / cust. int. saved
Option A and C: $30,000 / (360 + 240) = $ 50.0 / cust. int. saved
Option B and C: $30,000 / (420 + 420) = $ 35.7 / cust. int. saved

FIGURE 5.4
Recloser examples with customer interruptions saved.

On radial circuits with a major tee, the tee point is a prime candidate for a sectionalizing device on both branches off the tee point. Figure 5.4b shows an example comparing recloser options at a tee point; the most customer interruptions saved is for reclosers on both branches off the tee point. The best device (recloser or fuse) and the best location depend on the lengths, customer counts, and customer locations.

Figure 5.5 shows that applying a tap fuse is, not surprisingly, a very cost-effective way to reduce the number of customer interruptions. For longer

Fault rate = 0.2 faults/mile/year
1-phase fuse installation = \$400

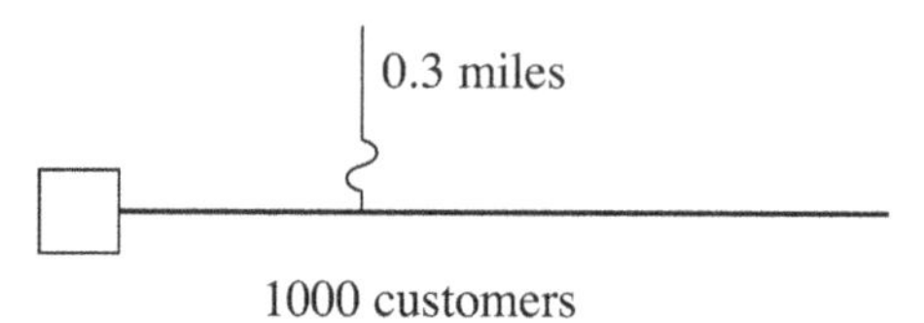

Customer-interruptions saved
= 1000 customers (0.3 miles) (0.2 faults/mile)
= 60 customer interruptions
Cost per customer interruption saved

$$= \frac{\$400}{60 \text{ cust. int. saved}} = \$6.7 \text{ / cust. int. saved}$$

FIGURE 5.5
Tap fuse example with customer interruptions saved.

taps, using a single-phase recloser rather than a fuse may be considered, and this question calls for a somewhat different analysis approach. One point to keep in mind is that most applications of sectionalizing equipment will not dramatically reduce utility restoration costs; it may take less time to patrol and find the damage, but the largest costs associated with restoration — getting the crews to the site and the repair time — do not change.

On overhead circuits, a recloser instead of a fuse will reduce the number of outage events by clearing temporary faults as a momentary interruption rather than having a blown fuse result in a sustained interruption. Consider a 2.5-mi tap with 50 customers. A fuse on that tap might blow (2.5 mi) (0.4 faults/mi/year) = one operation per year (assuming a fuse-blowing mode in which the fuse operates for temporary and permanent faults). Having a recloser on the tap may reduce the number of outage events on the tap by a factor of two (one operation every 2 years). If a single-phase recloser installation costs \$1500 more than a fuse installation and each fuse operation costs \$800 on average to replace, the recloser will pay for itself in 4 years. In this case, it is easier to estimate the impact by event, rather than using cost per customer interrupted.

To apply sectionalizing equipment, Duke Power uses the customer-interruptions-saved approach. The utility will implement projects in which the cost is less than \$50 per customer interruption saved using the cost guidelines in Table 5.3. These estimates are approximate and are intended to give engineers a "feel" for what is justified in sectionalization work. When protecting customers upstream from faults downstream, Duke Power assumes an annual permanent fault rate on overhead circuits of 0.2 faults/mi (0.32 faults/km) for circuit mainlines and 0.5 faults/mi (0.8 faults/km) for branch lines. Duke also considers a temporary fault rate that will manifest as a

TABLE 5.3

Duke Power Sectionalizing Cost Guidelines

Activity	Approximate Cost
Install one fuse	$200
Install three fuses	$500
Install three single-phase hydraulic reclosers	$7500
Install a three-phase vacuum recloser	$9500
Install a three-phase electronic recloser	$20,000
Relocate a recloser	$3750

sustained interruption unless a reclosing device or fuse saving is used. This temporary fault rate is needed to analyze the impact of a sectionalizing device to protect customers downstream from faults downstream, such as when considering the impact of fuse saving or considering whether a recloser is more appropriate than a fuse.

For mainlines or branch lines, Duke assumes a temporary fault rate of 0.2 for retrofitted lines (those in which transformer locations have been retrofitted with a local fuse along with other upgrades to reduce faults) and 0.8 for nonretrofitted lines. To estimate the impact of fuse saving using these temporary fault rates, consider the impact of each fuse downstream of the reclosing device. If the fuse coordinates with the upstream device (meaning the breaker or recloser can open before the fuse blows), multiply the length of that tap by the temporary fault rate on that tap to arrive at the number of customer interruptions saved with fuse saving. For a 0.5-mi long tap with 20 customers and a temporary fault rate of 0.8/mi/year, fuse saving will save eight customer interruptions annually on that tap (add up the remaining taps to evaluate the full impact).

Using Duke Power's numbers, the "value" of each sectionalizing device can be estimated by multiplying the fault rate by the value of each customer interruption ($50) by the exposure, or one can multiply the line exposure and the number of customers affected and then multiply by $10/customer-mile ($50/customer interruption × 0.2 faults/mi). Consider the example in Figure 5.6 when comparing two recloser scenarios. In this case, Duke can justify installing three single-phase reclosers at either location because both locations have a value of $9000, which is less than the installation cost of $7500. Installing reclosers at both locations is not justified because the value of the second recloser location is only $3000. It is not beneficial to move the reclosers from location B to A (or vice versa) because both have the same number of customer-interruptions saved.

Customer groupings are important to consider. Sometimes it makes sense to increase the mainline exposure to protect large customer groupings. For example, if a large grouping of customers is near the middle of a circuit, putting the recloser just beyond that grouping will improve reliability the most. Circuits with large groupings of customers near the end will have few opportunities for additional sectionalizing equipment unless automated

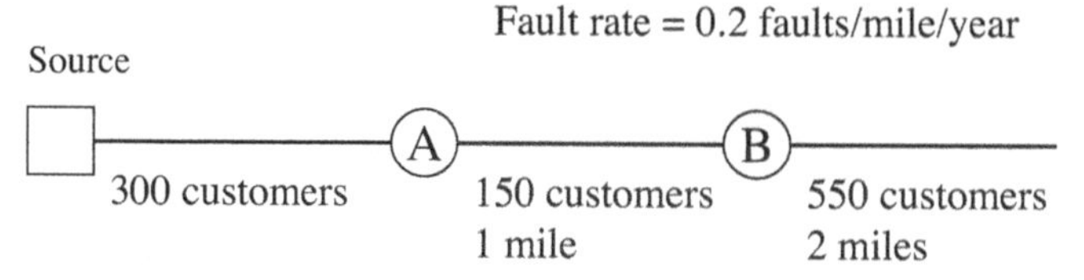

Customer exposure value = (0.2 faults/mi)($50/customer-interruption saved)
= $10/customer-mile

Value of each option:
A only: 300 cust (3 mi) ($10/customer-mile) = $9000
B only: 450 cust (2 mi) ($10/customer-mile) = $9000
A with B already there: 300 cust (1 mi) ($10/customer-mile) = $3000
B with A already there: 150 cust (2 mi) ($10/customer-mile) = $3000

FIGURE 5.6
Duke Power sectionalizing example.

tie-point equipment is used to transfer those customers to another circuit. For long, straight radial circuits, one recloser may be beneficial, but having a second recloser may be difficult to justify (although an automated tie recloser may be very beneficial).

For more advanced analysis, modules for reliability evaluations are available for most distribution analysis software, and some offer automatic optimization routines to help apply sectionalizing equipment. Distribution reliability software allows more fine-tuned estimates that can handle different fault rates and account for temporary and permanent faults; these can possibly account for the coordination between devices and for the differences between single-phase and three-phase devices.

5.2 Fault Sources

Faults are not evenly distributed along lines; they are not inevitable. Not all faults are "acts of God." Most are from specific deficiencies at specific structures. On overhead circuits, most faults result from inadequate clearances, inadequate insulation, old equipment, or trees or branches into a line. A first step in eliminating faults is to identify what is causing them. Keeping in mind that most faults result from specific structural deficiencies, field identification of fault sources is a key part of many of the construction-improvement programs that will be discussed. Field personnel can be trained to spot pole structures where faults have occurred or might be likely. Several common structural deficiencies include:

FIGURE 5.7
Fault arc damage on a gapped arrester. Courtesy of Duke Power.

- Poor jumper clearances
- Old equipment (such as expulsion arresters)
- Bushings or other equipment unprotected against animals
- Ground leads or grounded guys near phase conductors
- Poor clearances with polymer arresters
- Damaged insulators
- Damaged covered wire
- Danger trees or branches present

Consider Figure 5.7, which is an obvious example of a location where a fault has occurred — probably repeated faults. The arrester and pole are severely blackened from arc burn products, and the transformer bushing has cracked and is missing a piece. This structure has several severe deficiencies:

- The externally gapped arrester (which is possibly failed internally) provides a very short gap, easily flashed by birds or squirrels. In addition, it would take only a small lightning surge to flashover this structure.
- Neither the arrester nor the transformer bushing has animal protection.
- The guy wire attached near a phase conductor is uninsulated.

- The armless-type construction using post insulators provides little insulation on the pole.
- The equipment has no local fuse.

Overall, this structure is a disaster; it is highly susceptible to lightning, animals, small tree branches, and other debris. Fixing this pole requires a complete overhaul: replace the arrester, replace the transformer bushing, add animal guards, add a guy insulator, add a local fuse, and reframe with crossarms or other materials to get more insulation and electrical clearances. Although this example has many obvious problems, any one of the problems can be a source of faults, possibly repeated faults.

Inspections of structures to identify fault sources should be done from the ground (not while driving). Implement training out in the field for best results; show examples of fault sources. Walk the line and use binoculars; this is more effective than "riding the line." Some fault sources are not obvious and require looking at a structure from different angles.

When evaluating structures and possible fault causes, note the distinction between the cause of the fault and the deficiency. The cause of the fault may have been a squirrel, but the underlying source of the problem may have been poor electrical clearances (unprotected bushings, tight spacings, and so forth). Squirrels cannot be eliminated, but structures can be made more resistant to squirrel contacts.

5.2.1 Trees

Trees are major causes of interruptions and voltage sags. As discussed in Chapter 4, tree faults mainly occur from limbs or whole trees falling on circuits. Based on a sample of tree outages, Duke Power found that 73% of tree outages occurred when an entire tree fell on the line; 86% of these were from outside Duke's 30-ft right-of-way (Taylor, 2003). Dead limbs or trees caused 45% of tree outages. Also note that 25% of tree outages reported by crews were not caused by trees. For most utilities, tree growth accounts for less than 25% of faults.

Because most tree faults are from falling trees or limbs and because many of these are from dead or defective trees or limbs, utilities can reduce tree faults by implementing danger-tree programs to target defective trees and limbs. Dead trees are the most obvious candidates for hazard-tree removals. In a sample of permanent tree faults, Niagara Mohawk found 36% were from dead trees (Finch, 2001). Several utilities have implemented danger-tree or hazard-tree programs to remove defective trees, even if they are out of the normal trim zone or right-of-way. In an informal survey of seven utilities, Guggenmoos (2003) found that most utility hazard-tree programs removed about 5 trees per mile of circuit, with the most intense programs removing 10 to 15 trees per mile.

TABLE 5.4

Percentage of Tree Faults in Each Category

	Outage Events	CI	CMI
Tree Outside Right of Way (Fall/Lean On Primary)	26.0	37.2	42.5
Tree/Limb Growth	21.1	14.4	13.3
Limb Fell from Outside Right of Way	18.0	20.1	18.1
Tree Inside Right of Way (Fall/Lean On Primary)	12.6	14.8	15.2
Vines	10.0	3.6	3.1
Limb Fell from Inside Right of Way	8.7	9.8	7.5
Tree on Multiplex Cable or Open Wire Secondary	3.6	0.2	0.2

Data source: EPRI 1008506 [2005]
Southeastern US utility, 2003 – 2004
CI = Customer interruptions
CMI = Customer minutes of interruptions

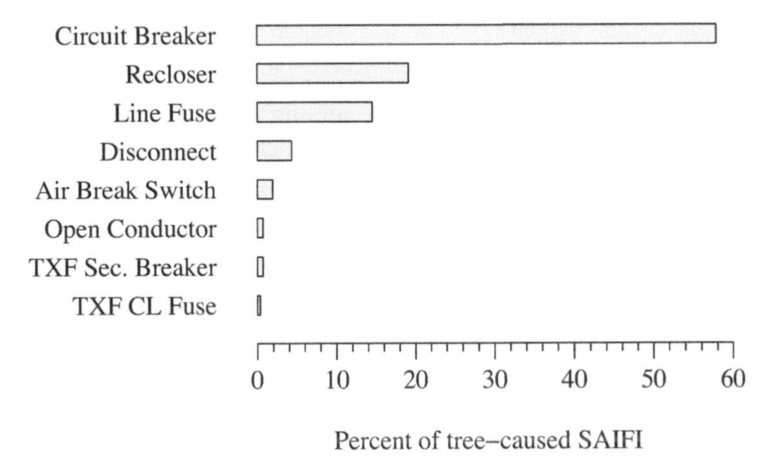

FIGURE 5.8
Tree-caused customer interruptions by interrupting device for one northeastern utility (major events included).

Better outage-cause codes can help utilities to target causes more precisely. Table 5.4 shows data for one utility for its breakdown of tree-faults and their impact on outages. Note that trees falling (whether from inside or outside the right-of-way) cause a much larger impact on the customer minutes of interruption relative to the actual number of outages. Likewise, vines and tree growth have relatively less impact on outage duration.

Interruptions on the mainline affect the most customers. Figure 5.8 shows the breakdown of tree-caused SAIFI for one northeastern US utility. This utility had almost 60% of tree customer interruptions from breaker lockouts. Duke Power (Chow and Taylor, 1993; Xu et al., 2003) also found that, of circuit breaker and recloser lockouts, trees caused 35 to 50% of the lockouts, which is over twice the rate of all tree outage events (trees cause 15 to 20% of all of Duke's outages).

Of tree-caused faults, a small number causes the greatest contribution to SAIFI and SAIDI. These are mainline faults. Restoration time is longest during major events, and tree failures normally have the longest (and most expensive) cleanup and repair. Finch (2003a) reported that, on Niagara Mohawk's system with over 2000 feeders in a given year, more than half of tree faults, half of customer outages, and half of customer hours of outage occurred on just 100 circuits (5% of feeders). Another utility in the Northeast shows similar trends in Figure 5.9; half of the customer outage hours were from outages on just 4% of feeders. These data included major events, which tend to exaggerate the "clustering" effect.

Goals and corporate structure of vegetation management also drive the effectiveness of programs. At some utilities, vegetation management groups do not really see their mission as improving reliability — their mission is to prune and/or remove trees. Utility goals often make the situation worse by giving the vegetation management personnel measures such as "cost per line mile" or "line miles trimmed per year." Furthermore, because the vegetation management group often has control over which circuits are cut, it could make its annual goals by trimming areas where the cost is low and the line miles are easy to get (pruning circuits near the tree crews' home bases, rather than where the bad circuits are). Such bias may result in making the goals, but not really addressing the circuits that need it. To avoid these problems, consider goals and tree-cutting standards carefully, and target sections of circuit that will have the most impact on reliability and power quality.

Targeting tree maintenance should help improve power quality and reliability and more efficiently manage tree-maintenance budgets. Target tree

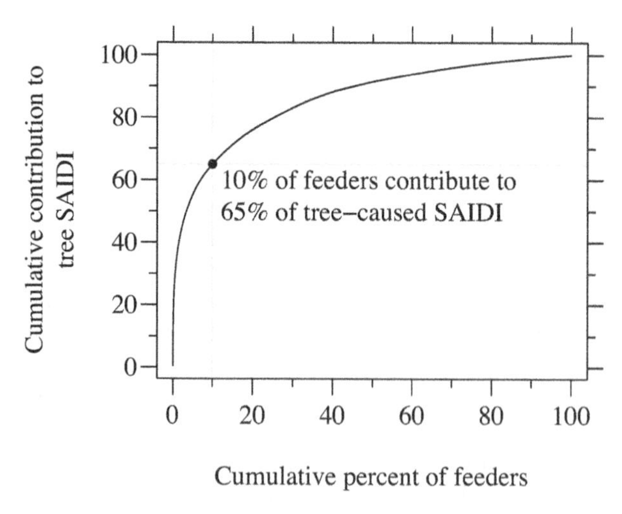

FIGURE 5.9
Allocation of tree-caused system SAIDI by feeder for one northeastern US utility.

maintenance to circuits with the most tree faults and target the most effective maintenance strategies. To do that, focus on the following:

- Mainline portions of circuits
- Circuits with more customers
- Circuits with a history of tree faults
- Circuits with higher voltage
- Hazard tree/branch removal
- Overhang removal

A number of utilities have begun using more targeted tree-maintenance programs. Most focus on removing danger trees and are targeted to mainlines and the worst performing circuits.

Eastern Utilities, a small utility in Massachusetts (now a part of National Grid), implemented a danger-tree mitigation project with the following characteristics (Simpson, 1997; Simpson and Van Bossuyt, 1996): (1) three-phase primary circuits were targeted; (2) dead or structurally unsound trees were removed; and (3) overhanging limbs were cut back. Trees were also "stormproof" pruned, meaning that they were pruned to remove less severe structural defects. This was mostly crown thinning or reducing the height of a tree to reduce the sail effect. On circuits where this was implemented, customer outage hours (SAIDI) due to tree faults were reduced by 20 to 30%. In addition, the program reduced tree-caused SAIDI by 62% per storm.

Eastern Utilities did not increase funding for its vegetation management program to fund the danger-tree mitigation project. Instead, it funded the program by changes to its normal vegetation management program. Crews did less trimming of growth beneath the lines. The company also embarked on a community communications effort to educate utility customers and win support for tree removal and more aggressive pruning. Additionally, Eastern Utilities did not remove viable trees without the landowner's consent. The company found significant overall savings from reduced hot spotting and an even more significant savings from reduced outage restoration costs.

After considerable study of tree-caused outages on its system, the Niagara Mohawk Power Corporation implemented a program called TORO (tree outage reduction operation) (EPRI 1008480, 2004; Finch, 2001, 2003a, b). As of 2002, based on 250 feeders completed, on 92% of the feeders, tree SAIFI improved an average of 67%. More recent results show even more improvement. The main aspects of the program are:

- Target work to the worst performing circuits based on specific tree-caused indicators.
- Remove hazard trees located on targeted circuit segments.
- Specify greater clearances and removed overhanding limbs on the backbone when possible.

- Lengthened tree-maintenance cycles on rural 5-kV systems from 6 years to 7 or 8 years. Urban and suburban systems are kept to a 5-year cycle.
- Look for opportunities to improve system protection; the utility recently added inspection for the presence of single-phase tap fuses.

Puget Sound Energy (PSE) implemented a hazard tree program called TreeWatch (Puget Sound Energy, 2003) that focused on removing dead, dying, and diseased trees from private property along PSE's distribution system. On the circuits where the program was implemented, the average number of tree-caused outages and average outage duration dropped as shown in Table 5.5. The company also found that it did not need to classify as many storms as major storms. Even in years with higher than normal average wind speeds, PSE declared fewer storms as major storm events (when 5% of PSE's electric customers are without power due to weather-related causes). Puget Sound Energy also estimates that it reduced the cost of tree maintenance on a per-circuit-mile basis by about 15%.

Several other utilities have also implemented targeted tree maintenance programs (EPRI 1008480, 2004; EPRI 1008506, 2005). Finch (2003b) provides details on programs that ECI helped implement, including those by Niagara Mohawk (discussed previously), Kansas City Power & Light (KCPL), and Flint EMC. KCPL and Flint EMC adjusted maintenance cycles to reduce cost and focus work on the most critical portions. On urban circuits, KCPL used a 4-year cycle on the backbone with a 2-year inspection to catch cycle busters and used a 5-year cycle on laterals. On rural circuits, KCPL used a 5-year cycle for all circuits. KCPL also developed a hazard-tree removal program based on results from its outage database. Flint EMC extended the maintenance cycle from 4 years to between 5 and 6 years on rural single-phase circuits.

Note that, although danger-tree programs and other targeted programs can improve power quality, they are not a panacea. Tree outages will still occur regularly. Many tree faults are from weather that causes tree failures of otherwise healthy trees. Danger-tree programs must be ongoing. As Guggenmoos (2003) shows in detail, with tree mortality rates on the order of 0.5 to 3% annually and the sheer number of trees within striking distance

TABLE 5.5

Results of the PSE TreeWatch Program

	Tree-Caused Outages			Average SAIDI, min/yr		
TreeWatch Year	Pre-TW	Post-TW	Change	Pre-TW	Post-TW	Change
2001 (82 Circuits)	272	170	37.4%	53.4	47.8	-10.4%
2000 (62 Circuits)	208	209	0.4%	100.0	86.7	-13.3%
1999 (26 Circuits)	241	172	28.6%	121.0	102.6	-15.2%

Data source: Puget Sound Energy [2003]

of distribution circuits, a danger-tree program cannot be a one-time expenditure.

Although better allocation of tree-maintenance programs with more targeted programs is the best way to target tree faults, other options exist. Covered conductors or spacer cable can help reduce the number of faults from branches causing conductor-to-conductor contacts. Duke Power has found that the best use of covered conductors is in areas where the trees are far above the three-phase lines (high overhang). Often, tree branches fall from the high canopy and land between two conductors. Covered conductors really help in this situation because no amount of tree pruning can eliminate the problem and the trees cannot be removed. Covered conductors can also help prevent midspan faults if the conductor clearances are not adequate.

However, consider that covered conductors have several drawbacks compared with bare conductors. Covered wire is much more susceptible to burndowns caused by fault arcs, especially in high-lightning areas. Covered conductors increase the installed cost somewhat. They are heavier and have a larger diameter, so the ice and wind loading is higher than a comparable bare conductor. The covering may be susceptible to degradation due to ultraviolet radiation, tracking, and mechanical effects that cause cracking. Covered conductors are more susceptible to corrosion, primarily from water.

Another approach to reducing the impact of tree damage is to coordinate the mechanical design so that, when tree and large limb failures occur, equipment fails in a manner that is easier for crews to repair. When a tree falls on a line, crews will have an easier repair if it breaks the conductors rather than breaking poles and other supports. The fault still occurs, but crews are able to move quickly to repair the damage and restore service.

BC Hydro (Kaempffer and Wong, 1996) has developed an approach to overhead structure design that considers the order of failure of equipment. Yu et al. (1995) developed methodologies for calculating conductor tensions under the stress of a large concentrated load (the falling tree or branch). BC Hydro used this analytical approach to analyze several of its standard designs and modify them so that the first components to fail were easier to repair. They found that trees falling near midspan and those falling near a pole were similar. For tangent structures, with #2 ACSR, the phase and neutral failed first when a tree fell on either conductor. For 336.4-kcmil ACSR, the pole tended to fail first. For angle structures, the guy grip for the phase and the tie wire for the neutral usually failed first and, for deadend structures, the guy grip tended to fail first.

5.2.2 Lightning

In the majority of cases, lightning causes temporary faults on distribution circuits; the lightning arcs externally across insulation, but neither the lightning nor the fault arc permanently damages any equipment. Normally, less than 20% of lightning strikes cause permanent damage. In Florida (high lightning), EPRI monitoring found permanent damage in 11% of circuit breaker operations that were coincident with lightning (EPRI TR-100218, 1991; Parrish, 1991). When equipment protection is done poorly, more equipment failures happen. Any failure in transformers, reclosers, cables, or other enclosed equipment causes permanent damage. One lightning flash may cause multiple flashovers and equipment failures. Even if a lightning-caused fault does no damage, a long-duration interruption occurs if the fault blows a fuse.

The protection strategy at most utilities is: (1) use surge arresters to protect transformers, cables, and other equipment susceptible to permanent damage from lightning; (2) use reclosing circuit breakers or reclosers to reenergize the circuit after a lightning-caused fault; and (3) maintain sufficient insulation to protect against induced voltages from nearby strikes. To improve lightning performance and arrester application, utilities can consider the following:

- *Insulation levels* — On overhead structures, review structure designs and field installations with the goal of having enough insulation to withstand most induced voltages from nearby strikes.
- *Arrester review* — Ensure that arresters are applied appropriately for adequate protection of equipment. Review arrester location and lead lengths to ensure that overhead and underground equipment are adequately protected. In some cases, it is also appropriate to use line-protection arresters to limit induced-voltage flashovers.
- *Arrester replacements* — Older arresters can fail, so by targeting these for replacement, the number of such faults will decrease.
- *Grounding* — Improving grounding can help reduce lightning damage on distribution circuits and reduce lightning surges into utility customers.

The main goal on structures is to maintain a sufficient insulation capability to avoid most induced voltage flashovers. The goal is a 300-kV critical flashover voltage. On wood structures, have at least 2.5 ft (0.75 m) of wood along all possible flashover paths. Consider the following advice to eliminate weak-link structures that may be particularly prone to flashover (see Figure 5.10):

- *Guy wires* — Use fiberglass guy-strain insulators rather than the small porcelain guy-strain insulators.
- *Ground wires* — Eliminate unneeded ground wires on poles near the primary.
- *Metal* — Avoid using conducting hardware when appropriate. Wood or fiberglass provides significant extra insulation.

FIGURE 5.10
Examples of weak-link pole structures. Courtesy of Duke Power.

For areas needing very high power quality, consider line protection as part of a hi-tech design. Although expensive, such designs can eliminate most lightning-caused faults, if they are designed with care. Options include line arresters and a well-grounded overhead shield wire.

To protect against direct strikes, a shield-wire system works by intercepting all lightning strokes and providing a path to ground. If the path to ground is not good enough, a voltage develops on the ground with respect to the phases (called a ground potential rise). If this is high enough, the phase can flashover (called a *backflashover*). Grounding and insulation are important. Good grounding reduces the ground potential rise, and extra insulation protects against backflashover. Because of the fast rise of the voltage surge from lightning, every pole structure must be grounded. Figure 5.11 shows an example of a shield wire design used by Duke Power, which grounds the shield wire at every pole, routes the downground wire away from phases to maximize insulation, uses fiberglass guy strain insulators, and attempts to achieve 25 Ω at each ground.

Arresters are normally used to protect equipment, but some utilities are also using them to protect lines against faults, interruptions, and voltage sags. To do this, arresters are mounted on poles and attached to each phase. For protection against direct strikes, arresters must be spaced at every pole (or possibly every other pole on structures with high insulation levels) (McDermott et al., 1994). This is a lot of arresters, and the cost prohibits

FIGURE 5.11
Duke Power's high-reliability design. Courtesy of Duke Power.

widespread use. The cost can only be justified for certain sections of line that affect important customers.

Arresters have been used at wider spacings such as every four to six poles by utilities in the southeast for several years. This grew out of some work done in the 1960s by a task force of eight utilities and the General Electric Company (1969a, b). Anecdotal reports suggest improvement, but little hard evidence is available. Recent field monitoring and modeling suggest that this should not be effective for direct strikes. One of the reasons that this may provide some improvement is that arresters at wider spacings improve protection against induced voltages. Nevertheless, applying arresters at a given spacing is not recommended as the first option; fixing insulation problems or selectively applying arresters at poles with poor insulation are better options for reducing induced-voltage flashovers. For direct strike protection, arresters are needed on all poles and on all phases. The amount of protection quickly drops if wider spacings are used.

One concern with arresters is that they may have a relatively high failure rate. Direct strikes can cause failures of nearby arresters. Something in the range of 5 to 30% of direct lightning strikes may cause an arrester failure. This is still an undecided (and controversial) subject within the industry. It is recommended that the largest block size available be used (heavy-duty or intermediate-class blocks) to reduce the probability of failure.

Field trials of arresters at various spacings on Long Island, New York, did not show particularly promising results for distribution line protection arresters (Short and Ammon, 1999). LILCO added line arresters to three circuits. One had spacings of 10 to 12 spans between arresters (1300 ft, 400 m), one had spacings of 5 to 6 spans (600 ft, 200 m), and one had arresters at every pole (130 ft, 40 m). Arresters were added on all three phases. Two

FIGURE 5-12
Lightning-caused fault on a Long Island Lighting Company 13-kV circuit. (From: Short, T. A. and Ammon, R. H., "Monitoring Results of the Effectiveness of Surge Arrester Spacings on Distribution Line Protection," *IEEE Transactions on Power Delivery*, vol. 14, no. 3, pp. 1142-50, July 1999. With permission. ©1999 IEEE.

other circuits were monitored for comparison. None of the three circuits with line arresters had dramatically better lightning-caused fault rates than the two circuits without arresters. Statistically, not much more than this can be inferred because the data are limited (always a problem with lightning studies). One of the most significant results was that the circuit with arresters on every pole had several lightning-caused interruptions, but, theoretically, it should have had none. Missing arresters are the most likely reason for most of the lightning-caused faults. One positive result was that the arrester failure rate was low on the circuit with arresters on every pole.

Automated camera systems captured a few direct flashes to the line during the LILCO study. Figure 5.12 shows a direct stroke almost right at a pole protected by arresters (arresters were every five to six spans on this circuit). Ideally, the arresters on that pole should divert the surge current to ground without a flashover. The arresters prevented a flashover on that pole, but two of the three insulators flashed over one pole span away. An arrester at the struck pole also failed.

Another field study showed more promise for line arresters. Commonwealth Edison added arresters to several rural, open feeders in Illinois (McDaniel, 2001). This utility uses an arrester spacing of 1200 ft (360 m); as a trial, it tightened the arrester spacing to 600 ft (180 m) on 30 feeders (and all existing arresters that were not metal oxide were replaced). The 30 feeders with the new spacing were compared to 30 other feeders that were left with the old standard. Over three lightning seasons of evaluation, the upgraded circuits showed that circuit interruptions improved 16% (at a 95% confidence level). Note that most of the interruptions were transformer fuse operations.

Arrester application is important. Arresters should have short lead lengths and have adequate animal protection; it is best to apply arresters downstream of a local fuse. Arrester reliability is also a concern. Replace older arresters whenever possible. Externally gapped arresters are a significant source of power quality problems; they can fail internally and cause repeated voltage sags and fuse operations.

Many older arrester technologies (see Figure 5.13 for examples) are still in place on distribution circuits. Many older distribution circuits have a variety of arrester technologies, including silicon-carbide, thyrite, pellet, and expulsion arresters. Older arrester technologies all have an internal gap. The gap can be a source of repeated power quality problems; it can spark over because of moisture. After the gap sparks over, "power follow" current will flow through the arrester; the amount of current depends on the material — anything from hundreds of amps in a silicon-carbide arrester to the bolted short-circuit level at the location in an expulsion arrester (more nonlinear arrester materials have less power follow current). The power follow current can blow fuses or trip other protective devices. Even on an arrester that is completely failed to a short circuit, the gap may be able to hold full system voltage until the gap sparks over.

The most prevalent type of in-service arrester is the gapped silicon-carbide arrester. Silicon carbide is a nonlinear resistive material, but it is not as nonlinear as metal oxide. It requires a gap to isolate the arrester under normal operating voltage. When an impulse sparks the gap, the resistance of the silicon carbide drops, conducting the impulse current to ground. With the gap sparked over, the arrester continues to conduct 100 to 300 A of power follow current until the gap clears. If the gap fails to clear, the arrester will fail. Annual failure rates have been about 1% with moisture ingress into the housing causing most failures of these arresters; an Ontario-Hydro survey found that 86% of failures were from moisture (Lat and Kortschinski, 1981). Moisture degrades the gap, damaging it outright or preventing it from clearing a surge properly. Darveniza et al. (1996) recommended that gapped silicon-carbide arresters older than 15 years be progressively replaced with metal-oxide arresters. Their examinations and tests found that a significant portion of silicon-carbide arresters had serious deterioration with a pronounced upturn after about 13 years. Externally gapped arresters

Silicon-carbide arrester Expulsion arrester Peroxide-pellet arrester

FIGURE 5.13
Examples of older arresters.

can especially be a source of repeated problems. If the arrester is failed, all it takes is an animal contact (or other debris) across a very short gap to cause a fault on the system.

Use every opportunity to replace older arresters. Doing so will remove threats to power quality from repeated flashovers and improve the protective level provided by modern arresters. With poor grounding, lightning strikes to distribution line subject more equipment to possibly damaging surges due to ground potential rise. Lightning current must flow to ground somewhere; if the pole ground near the strike point has high impedance, more of the surge flows in the line, exposing more equipment to possible overvoltages. At the point of a direct stroke to a distribution conductor, the huge voltages flash over the insulation, shorting the phase to the neutral. From there, the lightning travels to the closest ground; if no proper pole ground is nearby, the next most likely path is down a guy wire. If the path to ground is poor, the phase conductor and neutral wire all rise up in voltage together. This surge (on the phase and the neutral) travels down the circuit. When the surge reaches another ground point, current drains off the neutral wire, which increases the voltage between the phase and the neutral. The voltage across the insulation grows until a surge arrester conducts more current to ground or another insulation flashover occurs. Figure 5.14 shows a drawing describing the ground potential effect.

Grounding improvement programs can also improve reliability and power quality. Good grounding helps confine possible lightning damage to the immediate vicinity of the strike. If the distribution insulation flashes over, the short circuit acts as an arrester and helps protect other equipment on the circuit (as long as grounds are good). Normally, the ground resistance right at a piece of equipment protected with an arrester does not have an impact on the primary-side protection. The primary surge arrester is between the phase and ground. For a lightning hit to the primary, the arrester conducts the current to ground and limits the voltage across the equipment insulation (even though the potential of all conductors may rise with poor grounding).

Grounding does play a role for transformers because they are vulnerable to surge current entering through the secondary side. Poor grounding pushes more current into the transformer on the secondary side and increases the possibility of failure. Poor grounding also forces more current to flow into the secondary system to houses connected to the transformer. Additionally, poor grounding may also push more current into telephone and cable television systems.

One difficulty with lightning-improvement programs is predicting their effectiveness. This is true for grounding-improvement programs, arrester application or replacement programs, and insulation improvements. Normally, utilities estimate a certain percentage reduction in the number of lightning-caused interruptions, but it is an uncertain science.

5.2.3 Animals

Faults caused by animals are often the number-two cause of interruptions for utilities (after trees). An animal that bridges the gap between an energized

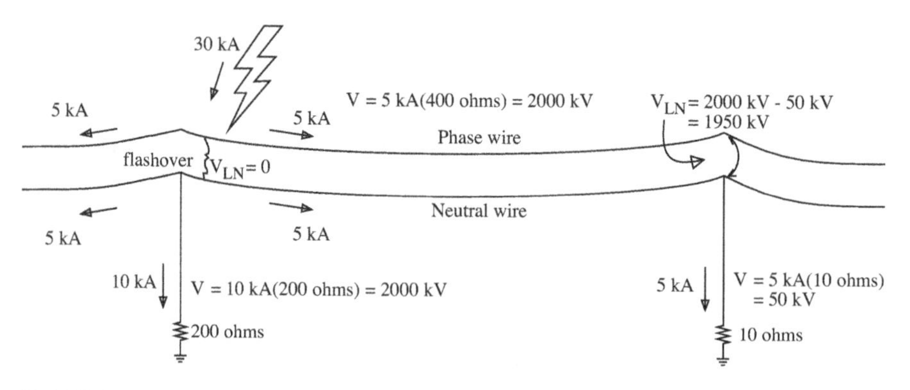

FIGURE 5-14
Impact of grounding.

conductor and ground or another energized phase will create a highly ionized, low-impedance fault-current path. The fault will cause a sag in voltage to nearby customers and an interruption to the portion of the circuit covered by the upstream protective device. An animal can cause a temporary or a permanent fault. Animal-caused faults are normally phase to ground.

Properly applied bushing guards in conjunction with covered jumper wires can effectively prevent most animal-caused faults. For example, in one Duke Power circuit, animal guard application on all distribution transformers on the circuit reduced animal faults from 12 per year to an average of 1.5 per year (Chow and Taylor, 1995). In Lincoln, Nebraska, application of animal guards on all 13,000 transformers reduced animal-caused faults by 78% (Hamilton et al., 1989).

An EPRI survey (EPRI TE-114915, 1999) points out where most animal-caused outages occur — at equipment poles, where phase-to-ground spacings are tight (see Figure 5.15). Transformers are the most widely found pole-mounted equipment on a distribution system. Unprotected transformer poles normally have many locations susceptible to animal faults: across the transformer bushing, across a surge arrester, and from a jumper to the transformer

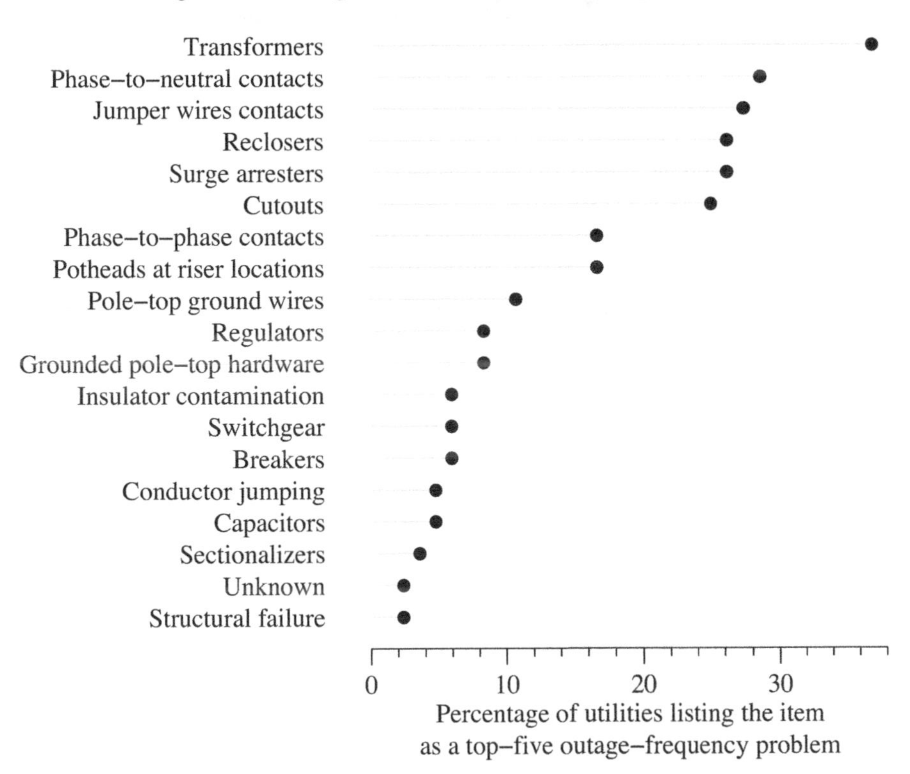

FIGURE 5.15
Overhead distribution points most susceptible to animal-caused faults. Data source: EPRI TE–114915 (1999).

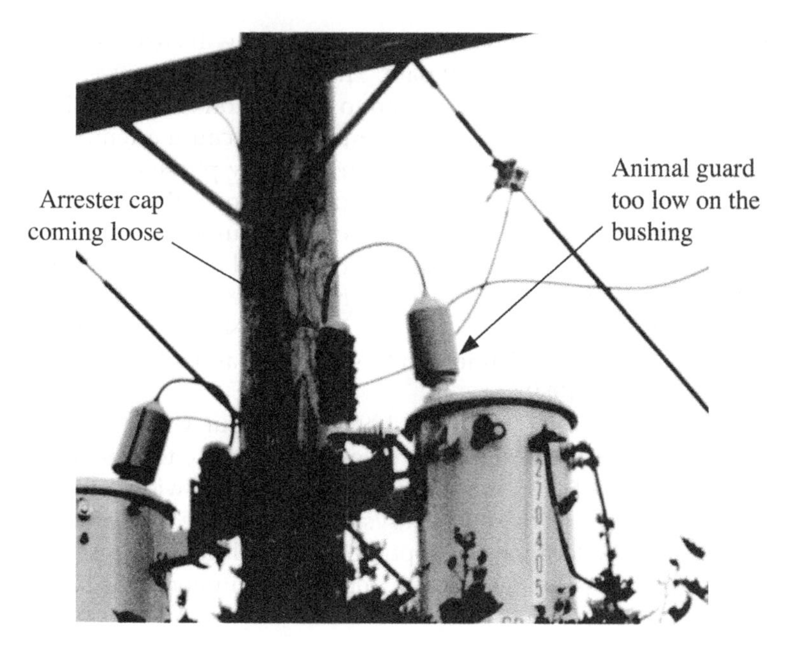

FIGURE 5.16
Examples of poor animal guard installations.

tank. Riser pole installations, regulators, reclosers, and capacitor banks all have susceptible flashover paths unless utilities employ protective measures. Poles without equipment are much less susceptible to animal faults (exceptions are poles with grounded guys or ground wires near phase conductors). This indicates where to concentrate programs to reduce animal-caused faults.

Choose and specify animal guards appropriately. Use an animal guard that is large enough to cover the top of the bushing adequately. Do not just specify and purchase equipment (transformers, arresters, and so on) with "animal protection." A specification of animal protection can lead to manufacturers applying the cheapest animal guards, which may be flimsy or undersized. Proper installation is important for bushing guards. The guard should be placed securely around the bushing and locked into place according to the manufacturer's directions. Height placement is also important (see Figure 5.16). For bushing protectors, crews should leave some room between the bottom of the bushing protector and the tank. Animal guards are not fully rated insulators; they can track and flash over, so it is not desirable for the animal guard to bypass the insulator fully.

The most common mistake is to leave some energized pathways exposed. Crews should cover every bushing and arrester and use insulated leads on all jumpers. In a survey on 253 poles that should have had animal protection, Pacific Gas & Electric (PG&E) found that 165 (65%) were found with incomplete or improperly installed devices (California Energy Commission, 1999). The most common problems were missing bushing covers or missing covers on jumpers (see Figure 5.17).

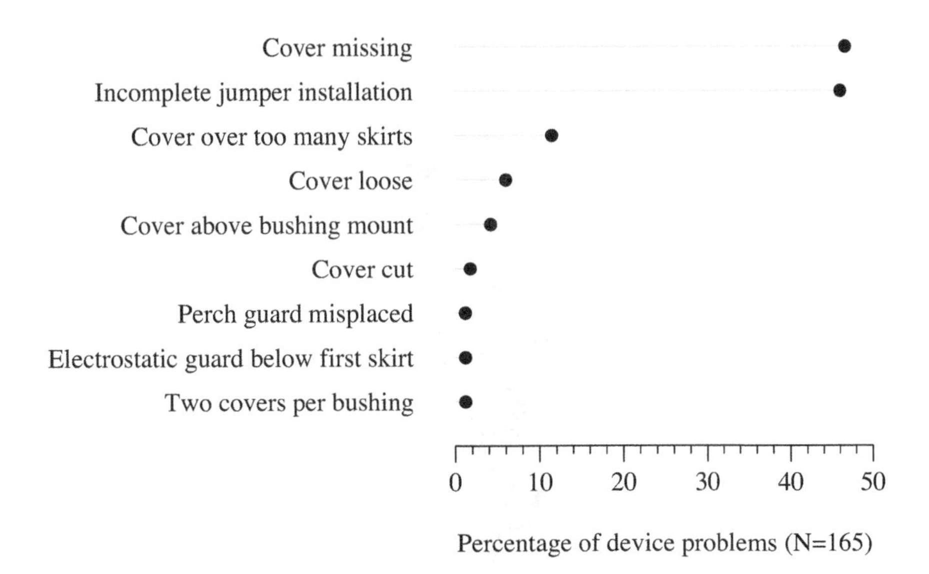

FIGURE 5.17
Problems noted at poles with missing or improperly installed animal protective devices. Data source: California Energy Commission (1999).

Construction practices that can increase animal faults include ground wires and grounded guy wires near the primary. Figure 5.18 shows an example of an installation that could have repeated animal faults because of the short phase-to-ground separation. The best solution is to remove the ground wire. When possible, separate energized conductors from grounded conductors and other phases. Separations of at least 18 to 24 in. (46 to 61 cm) are normally sufficient; spacings less than 10 in. (25 cm) are very prone to animal faults. When sufficient air clearances are impossible, use insulated phase and/or ground conductors. At riser poles and some other equipment structures, insulated supports provide better animal protection than bonded metallic supports (as in Figure 5.19), and a variety of wood or fiberglass alternatives is available.

5.2.4 Cable and Equipment Failures

Programs to reduce equipment failures center around identifying the most problematic equipment and replacing it on the most critical portions of circuits. For many utilities, cables and splices contribute to a large portion of equipment failures that are particularly costly to replace or repair.

Many utilities regularly replace cable. Program policies are done based on the number of failures (the most common approach), cable inspection, customer complaints, or cable testing. High molecular weight polyethylene and older XLPE are the most likely candidates for replacement. Most commonly, utilities replace cable after two or three electrical failures within a given time

FIGURE 5.18
Electrocuted cat and the "problem" ground wire. Courtesy of Duke Power.

period (see Table 5.6). Several utilities have programs specifically to replace substation exit cables. Substation exits are critical circuit components; faults there interrupt many customers, the fault is high magnitude, and the voltage sags to almost zero.

Another way to learn more about cable and other equipment failures is to do sample surveys. Sample surveys allow one to learn more about equipment failures than can be learned from outage codes. If cable splice failures seem to be problematic, start retrieving failed samples from the field. On the samples received, track the manufacturer and the type of splice and compare this against estimates of populations by splice type. This may highlight splices of a certain type that are failing at a higher rate than others.

With the field-failed samples, postmortem teardowns may reveal modes of failure and identify critical modes of degradation: Was water entry a key? Are signs of predischarges or other arcing present? Are breakdowns from mechanical or electrical stress visible? The teardowns might also reveal workmanship issues or application issues. The failed samples can be used

FIGURE 5.19
Riser pole with a metallic support that is highly susceptible to animal faults. Courtesy of Duke Power.

TABLE 5.6
Typical Cable Replacement Criteria

Replacement criteria	Responses (n=51)
One failure	2%
Two failures	31%
Three failures	41%
Four failures	4%
Five failures	6%
Based on evaluation procedures	16%

Data source: Tyner [1998].

along with information on the locations of failures to test various hypotheses about the failures. Do the failing type-X splices happen to be in locations where loading is heavy? Are they in especially wet locations? Confirmed hypotheses point the direction to remedies, including replacement. Equipment teardowns can be used on splices, cables, arresters, reclosers, and many other pieces of distribution equipment.

Equipment purchasing is another time to review equipment carefully to ensure that quality equipment will be purchased. Some equipment is better than others. Tear equipment samples apart. Check for design weaknesses and poor workmanship. Are seals sufficient to keep water out? Are materials susceptible to UV degradation (for polymer materials)?

5.3 Programs to Reduce Fault Rates

The best way to reduce faults over time is to "institutionalize" fault-reduction practices. After identifying the most common fault sources, implement programs to address them so that performance improves continually. Options for such programs include:

- *Design review* — The first step in implementing fault-resistant designs is to start with good designs.
- *Outage follow-ups* — On-site outage reviews can help identify weak points and reduce fault rates. Faults tend to repeat at the same locations and follow patterns. By identifying the location of the fault and any structural deficiencies that contributed to the fault, the deficiencies can be corrected to prevent repeated faults in the future.
- *Construction audits* — Construction audits help reinforce practices to ensure that crews build to the specifications, and audits help educate crews on why things are done and how the designs minimize faults.
- *Problem-circuit audits* — Just as pole locations can have repeated faults, faults can cluster on some circuits. The goals of a problem-circuit audit are to (1) identify the deficiencies that are the most probable sources of faults; and (2) correct deficiencies to avoid repeat fault events.
- *Upgrade and maintenance projects* — Utilities can shore up weak areas by any one of several construction upgrade programs: animal-guard implementations, arrester replacements or new applications, and cable-replacement programs. Maintenance projects include tree clearance, danger tree programs, and pole inspection and replacements.

There are no magic programs or devices or quick fixes. Utilities must maintain consistency. For the most part, these are not 1- or 2-year programs. Maintaining fault resistance must be an ongoing process that becomes a core part of utility operations.

Duke Power pioneered the strategy of identifying and removing sources of faults on the system. By doing this, the utility is able to maintain very respectable reliability numbers despite the fact that its service territory has regular severe weather and that Duke Power is mainly an overhead utility with predominantly all-radial systems. Since first starting to implement fault reduction programs, Duke has reduced its SAIFI by 20% during a 10-year period (Figure 5.20).

Because fault sources are most often at equipment poles, consider programs targeted specifically at equipment poles. Duke Power has many single-phase CSP transformers with an old arrester that can fail at any time, minimal animal protection, and no local fusing. These locations are

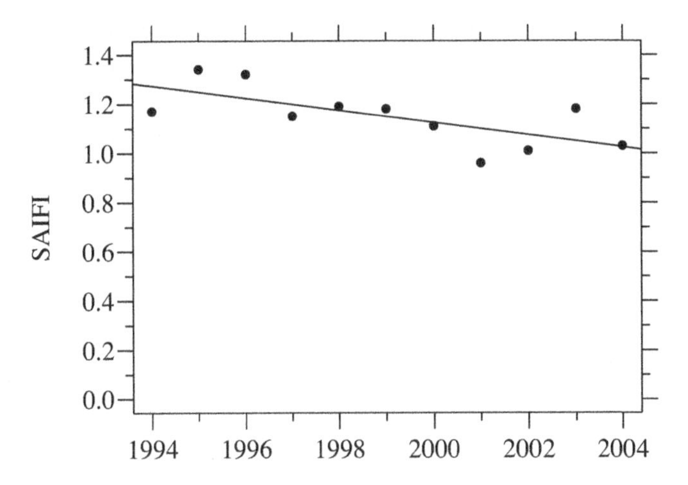

FIGURE 5.20
Duke system SAIFI with trendline.

tied for second place in being the largest contributor of outage minutes on Duke's system. To address this problem area, Duke has a transformer retrofit program in which a crew will install a local cutout with a surge-resistant fuse, a new lightning arrester, insulated leads, and animal guards. While a crew is set up at a pole, they also insulate uninsulated primary guys and remove pole grounds above the tank level. Duke applies this program on a circuit-by-circuit basis, with each region developing a prioritization plan based on outage database records.

5.4 Outage Follow-Ups

On-site outage reviews can help improve outage-cause codes, identify weak points, and reduce fault rates. Faults tend to repeat at the same locations and follow patterns (Chow and Taylor, 1993). Consider animal faults — one particular pole, which happens to be a good travel path for squirrels, may have a transformer with no animal guards. The same location may have repeated faults. By identifying the location of the fault and any structural deficiencies that contributed to it, the deficiencies can be corrected to prevent repeated faults in the future.

The main goals of an outage-review program are:

- *Outage database improvement* — Did crews enter the fault code or outage code correctly based on available information? If not, feedback can correct the mistake and lessons learned can help prevent future mistakes.

- *Training* — Field engineers and operating crews get more on-the-job training about specific design and construction deficiencies that lead to faults.
- *Source* — Identify the most probable cause of the fault (tree limb, animal, lightning, etc.)
- *Corrective action* — Construction deficiencies should be corrected to avoid repeat fault events. Corrective action is a prime goal of an outage review.

During an outage review, the main tasks of the reviewer are to review the data at hand and address the following questions:

- What was the most likely location of the fault? What was the flash-over path?
- What was the most likely cause of the fault?
- Was the outage code entered correctly?
- Was the faulted circuit adequately covered by protective devices? Was the fault on an unfused tap?
- Did protective devices operate as expected?
- What structural deficiencies contributed to the fault?
- Are the deficiencies likely to lead to additional faults?
- How can the deficiencies be corrected?

Based on addressing these issues, corrective action can be initiated if it is warranted. Corrective action could include rebuilds on one or more structures and could include tree/branch clearance. Corrective action could also include fixing unrepaired damage related to the fault.

The first step is to find the most likely location of the fault that caused the outage. Permanent faults are relatively easy to pinpoint because the crew had to fix damage at one or more locations. Temporary faults are harder to pinpoint. If a fuse blows, the area narrows considerably. For areas with repeat fuse operations, careful patrols may identify areas where repeated faults occur. Still, if a tap fuse operates but is refused successfully, the cause may have been a squirrel across a bushing, a tree branch that fell onto then burned off of a line, or wind pushing two conductors together. It often takes a trained eye to determine the cause.

Many outages are classified as "unknown." For some outages, the cause will remain unknown. Although the exact cause may remain unknown, information about the outage can help point the way to finding deficiencies. The time of day and the weather during the fault can help suggest a cause. Clear weather may suggest an animal fault, especially if it happens during the daytime. Stormy weather suggests lightning. An outage history for the protective device may reveal patterns. The probable cause can help the outage investigator look for deficiencies that could have led to the cause. Several options are available to decide which outages to review:

- All outages that affect over 500 customers (or some other number)
- All mainline outages
- All outages on problem circuits
- All outages on critical-customer circuits
- Outages on protective devices with excessive operations over a given time period (pick a threshold based on whether the device is a fuse, recloser, or circuit breaker)
- A random portion of all outages

Utilities could also use combinations of these options in forming criteria for outage reviews. Responsibility for outage follow-ups will normally fall to a local field engineer.

In addition to regular outage reviews, a utility may audit a sample of the outage reviews. A "reliability auditor" could visit the outage location with the local field engineer. The main point of such an audit is to educate and reinforce consistent approaches to outage follow-ups. If, at the time of the audit, the corrective action has been done, the auditor and the local field engineer can review that action to determine whether it was done properly. For example, check that animal guards were attached correctly and that sufficient clearances were maintained.

5.5 Problem-Circuit Audits

Just as pole locations can have repeated faults, faults can cluster on some circuits. These faults may be from consistently poor construction on a circuit, heavy tree exposure, or a few poor structures with repeated faults. The goals of a problem-circuit audit are:

- *Source* — Identify the deficiencies that are the most probable sources of faults.
- *Corrective action* — Correct construction deficiencies to avoid repeat fault events.

Criteria for designating problem circuits can come from many options: circuits with critical customers that are having problems, circuits with excessive momentary interruptions, circuits with long-duration interruption problems (high SAIFI or SAIDI), or circuits with excessive customer complaints.

The problem should also direct the focus of the audit. If it is momentary interruptions, review the section covered by the circuit breaker or recloser that is excessively operating. Pay special attention to animal and lightning susceptibility on the circuit. If the problem is long-duration interruptions, concentrate circuit patrols and reviews on the feeder backbone. Pay special attention to susceptibility to trees; a review of the outage records for the circuit should reveal even more information on where to look for problems. Some specific items for a repair checklist could include:

- Unfused tap lines
- Animal guards missing on any surge arresters or bushings
- Transformers without local, external fuses
- Guy wires without proper insulation
- Pole ground wires run near the primary
- Old arresters
- Heavy tree coverage
- Danger trees
- Overhanging dead limbs
- Damaged insulators or bushings
- Deteriorated or damaged equipment
- Poor clearances
- Vines on equipment poles
- Midspan conductor faults (poor clearances)
- Wire too small for fault duty, including the neutral (through faults)
- Covered conductors subject to burndowns (arcing faults)

In addition to construction-related issues, a utility should also review the protection schemes on a circuit and review other ways to reduce the impacts of faults (EPRI 1001665, 2003). Are reclosers and other devices coordinated properly? Does the circuit have enough protective devices? Are there any unfused taps? Are all mainline equipment structures protected by a local fuse? Is reclosing done appropriately for that circuit? Are circuit breaker relays set fast enough to protect conductors from burndown?

5.6 Construction Upgrade Programs

Outage follow-ups and construction audits along with reviews of outage data help pinpoint the weaknesses in existing construction. Utilities can shore up these weak areas by any one of several construction upgrade programs:

- Animal-guard implementations
- Arrester replacements or new applications
- Tree clearance
- Danger tree programs
- Pole inspection and replacements
- Cable-replacement programs

Other options are possible, depending on past construction deficiencies. For example, if a utility has at some point built a number of lines with a shield wire (overhead neutral) that perform poorly, a program might target them for replacement or upgrades. A shield wire design can perform well if it has good insulation and good grounding, but it can also perform poorly if not done right; the grounding downlead can reduce insulation levels and make lightning contacts more likely and can increase the likelihood of animal contacts and contacts from other debris. Then, a shield-wire upgrade program could target these sections for replacement or for insulation upgrades to reduce animal and lightning faults.

Equipment replacement programs also fall under the upgrade category. Many utilities have cable replacement programs. Program policies are done based on the number of failures (the most common approach), cable inspection, customer complaints, or cable testing.

Because fault sources are most often at equipment poles, consider programs targeted specifically at these poles. For example, a transformer retrofit program could upgrade all transformer locations to the utility's design specifications. This could include replacement of old arresters with tank-mounted polymer metal-oxide arresters, animal guards on bushings and arresters, covered jumper leads, and elimination of grounded wires above the transformer tank (including grounded guy wires).

It costs much less to upgrade construction when a crew is already set up at a pole. Consider implementing procedures and checklists for the crew to fix problems on the pole *whenever* they are set up for work on one. Options to consider in such a program are:

- Add animal guards on unprotected bushings or surge arresters.
- Replace old surge arresters
- Add fiberglass guy insulators to uninsulated guys near the primary.
- Strip out unnecessary grounds above the neutral wire (especially those near primary conductors or jumpers).
- Replace any flashed or damaged equipment.
- Install a local fuse on CSP transformers.

5.7 Using Outage Databases

On a final note, these strategies and programs for improving reliability and power quality should be targeted for maximum benefit. Outage databases can help identify problems leading to faults. They can help identify which circuits to target, which areas have the most problems with trees, which areas have the most problems with animals, and so on. Outage databases can also help judge the effectiveness of improvement programs.

Without accurate and thorough outage databases, there is no evidence to justify reliability improvement programs. The best strategy is to provide relatively simple database entry codes and remarks fields. Guidelines and training for entering the outage data should be provided. Ensure that the codes are entered accurately by using audits, assessments, and budgetary motivation. Regional budgets for reliability improvement programs are often based on the reliability information extracted from the outage database. Examples of reliability-based budget allocations include underground cable replacement, vegetation maintenance, and "worst circuit" programs. If the correct failure codes are not used in the outage database, a region may lose money for certain reliability and asset management programs.

All outages contain an element of mystery. The initial coding of outages is based on the best judgment and training of the "first responder" and the employee who enters the data. However, the initial coding often does not identify the primary root cause. One of the major goals of an outage follow-up process is to identify this cause. Most outages have more than one root cause. The primary one is the root cause for which a utility can implement economical corrections within a reasonable period of time. The purpose of codes and coding conventions is to identify the primary root cause of each outage.

Most outage databases provide standard information about the outage such as date, time, customers affected, and other information. For reliability analysis, however, the code fields and remarks field provide information critical to modern asset management strategies.

Having accurate and precise outage code systems increases the usefulness of outage databases. *Weather* is a common outage cause code, but what does that mean? If a tree knocks down a distribution line during a storm, is that weather? How does one differentiate between lightning and tree-caused outages during a thunderstorm? Another common blunder that crews make is to tag a cause as "cutout" when the cause was really something downstream of the cutout and the fuse operated to clear the fault (the cutout operated properly). As much as possible, good outage code systems should separate the root cause of the outage from the weather, the protective device that operated, and the equipment affected. Use a separate category for weather (and indicate a major storm separately). Also, try to have codes that reveal deficiencies; these may include inadequate clearances, deteriorated equipment, missing animal protection, or low insulation.

Duke Power's system (other utilities use similar systems) provides a good way to characterize an outage by specifying four code fields for an outage: (1) interrupting device; (2) cause or failure mode; (3) equipment code; and (4) weather. Used in combination with each other, these codes are quite accurate in describing what is known about the cause of an outage or, in some cases, what is not known about an outage. In more detail, these codes mean:

- *Interrupting device* — Fuse, line breaker (recloser), station breaker, transformer fuse, etc.
- *Cause or failure mode* — Animal, tree, unknown, etc. If the "cause" is not known or does not fall into a predetermined cause category, the cause code becomes the failure mode. Often, the failure mode is the only information available. Examples are "burned," "broken," "malfunctioning," and "decayed." Failure modes imply equipment failures, so an equipment code is usually a required entry if using a failure mode. Failure modes represent partial information about the cause, in that the root cause may remain unknown but certain things about the failure are known. For example, if a station breaker fails to trip, a specific failure mode called "fail to interrupt" may have occurred. For the equipment code (see later), enter the code for "station breaker." This system is flexible. The same failure mode can be used with other equipment codes to describe other devices that "fail to interrupt." If there is no specific failure mode, a generic failure mode such as "malfunctioning" or even "broken" can be used.
- *Equipment code* — This is for equipment failures, to specify the equipment that failed.
- *Weather* — It is useful to know whether an equipment failure or unknown interrupting device outage occurred during lightning or other stormy weather. Animal outages normally occur in fair weather. By studying combinations of time-of-day, weather, and interruption device activity on a utility, it is often possible to classify unknown outages into likely cause categories. For example, on the Duke Power system, an unknown outage on a tap fuse on a spring morning in fair weather is highly likely to be an animal outage.

Often, discussion concerns if weather is an outage cause or a contributing factor. From a reliability perspective, the electric distribution system is designed to withstand moderate to bad weather conditions. As a primary root cause, weather that exceeds the design parameters of the system can be considered a primary root cause. However, these weather causes are the exception. Weather situations that may be considered primary root causes include: lightning, wind loading exceeding design parameters (tornadoes), ice loading exceeding design parameters (e.g., 3 in. of ice), or flooding. Even given that, utilities should avoid declaring stormy weather such as wind or lightning as a cause. Evidence (burn marks, for example) may show that lightning caused an arrester failure, but if evidence is not available, it is better to code the cause as an equipment failure during a lightning storm. Do not use weather as an excuse to "explain away" outages.

Reliability programs can prevent many outages during stormy weather. Properly funded and executed vegetation management programs can cut down on outages during wind. Properly designed overhead lines and equipment protection can substantially reduce outages during lightning. If the utility is not attempting to make the system more outage resistant during stormy weather, it may be thinking of weather as a cause, rather than as a contributing factor.

Engineers often want to specify additional codes such as vintage year (for equipment failures), overhead/underground indicators, how the outage was restored, and others. Most of these codes can be incorporated in the four main codes given previously. First responders will likely ignore the rest. For example, underground cable failures have equipment codes that are obviously underground types. Experience on the Duke Power system has shown that the four codes mentioned earlier are about the limit that a first responder can accurately handle, even if other codes are requested.

Have a generic comment or remarks section to note specifics that might not be clear from the outage codes. This helps with later analysis and allows for keyword searching revealing patterns. The remarks field contains vast amounts of information if used liberally. Provide at least 256 characters in the remarks field. Encourage first responders and dispatchers to use the remarks field to describe what happened during the outage. Modern database applications can quickly search and count outages that contain certain keywords or phases within the remarks field. These database applications can also filter out records that contain certain keywords. Here are some examples used at Duke Power:

- The code system had no code to differentiate between live or dead tree outages. However, the crews almost always noted in the remarks field whether the tree or limbs involved were "dead" or "rotten." This method of counting dead tree outages vs. live tree outages was found to be a highly accurate method to determine whether the annual danger-tree survey and removal was effective.
- One hypothesis was that 3D or smaller transformer fuses were more susceptible to nuisance fuse outages during lightning than 5D or larger transformer fuses. There were a large number of outage records of fuse outages on transformers during lightning, but the fuse size was not a record that anyone kept. However, it was discovered that the crew almost always called in the fuse size when they replaced a fuse. The dispatchers put this information in the remarks field. A contextual database search on the remarks field of

transformer outages found more than 7000 cases per year in which the transformer fuse size was supplied. This sample size was sufficient to allow normalization of 3D and 5D transformer outages against the total population of transformers that would have such fuses. The result of this analysis showed no significant difference in the performance of these two fuses.

- There was a concern that squirrels were causing significant damage to overhead conductors, connectors, and equipment by literally biting and gnawing on various items. The remarks field was searched to find how widespread this behavior was. Duke Power determined that squirrels will chew and gnaw almost anything, but bare aluminum conductor is their favorite. Also, although widespread, this phenomenon is not a significant reliability problem.

When outage code systems are designed or modified, it is necessary to balance the desire for as much information as possible with the reality that overloading crews with too many options is counterproductive. Stick with clear choices and do not overwhelm crews with options. Do not use codes that are too generic to provide useful information. Do not use codes that are too complicated for first responders to understand. In either case, little useful information is obtained. A balance must be maintained between being too generic and too complicated. Here are some additional guidelines for outage coding:

- *Subcodes* — When appropriate, subcodes provide extra information. As an example, it may make sense to have a code or subcode to denote whether the structure had an animal guard. If that adds too much complexity, have crews enter a generic comment to indicate the structural deficiency (the missing animal guard). Also, for animal outages, crews should only mark "animal" if they find the remains of the animal or other proof.
- *Unknown* — Using unknown as a cause is better than guessing. A wrong cause code can result in resources directed to the wrong problem.
- *Equipment failures* —Equipment failures are prime candidates for fine-tuning. For example, a subcode could indicate the equipment construction or vintage or even manufacturer. For cables, and especially for splices, knowing the manufacturer and vintage can help determine targeted replacement programs. The mode of failure is also an important consideration: was the equipment broken? Was the failure from decay?

Consider an example outage code for a tree contact during a storm that caused a downed wire that might be tagged as follows:

Interrupting device: breaker

Cause: vegetation (dead tree)

Equipment: bare conductor (downed wire)

Weather: wind/rain

This clearly separates the cause (the tree) from the impact on the system (downed wires) and notes the weather conditions during the event. The items in brackets denote subcodes that add useful information; in this case, the subcodes reveal that the tree was dead (points to a better hazard-tree program) and that it caused a downed wire.

An arrester failure found during a lightning storm might be classified as:

Interrupting device: transformer fuse

Cause: equipment failure

Equipment: arrester (catastrophic)

Weather: lightning storm

In this case, the cause of the arrester failure is not known; it was probably lightning, but that is not known for sure. It is not even known for sure whether the arrester failed during the storm (the arrester may have failed earlier and was only found when crews arrived at the scene). However, because the arrester obviously failed just downstream of the fuse and the transformer was still operational, the arrester is very likely to have caused the fault.

Consider a squirrel outage across a bushing:

Interrupting device: tap fuse

Cause: animal

Equipment: bushing (no animal guard)

Weather: clear

The use of a subcode for bushing that highlights a common deficiency (no animal guard) helps direct resources to repair the deficiency.

Consider also the way in which crews provide the information. Using mobile computers allows the most direct data entry, but crews are prone to using the software incorrectly (not using it or entering the data to process the menus as quickly as possible). Training and straightforward user interfaces can help. The advantage of computer data entry is that the form can adapt to the scenario at hand and fill in data from the outage management system (like outage start time). Having a crew call back outage information to dispatchers allows the dispatchers to query the crew to make sure that the codes are entered consistently. Paper data entry sheets can be relatively easy for crews to interpret, but they limit flexibility for subcodes and options.

Utility culture should also encourage accurate outage codes. Provide a document to show how all outage codes are supposed to be used and provide training. These guidelines and "code sheets" should be provided to everyone who has anything to do with entering outage codes. At Duke Power, dispatchers enter outage codes called in by first responders. At the 24-hour center, dispatchers are graded on how accurately the code guidelines are followed. Local reliability engineers or technicians are the "code police." Each day, these employees review all the outage records in their location for the past 24 hours or over the weekend. They make sure that the code guidelines are followed. If something is missing, unclear, or contradictory about the information in the outage record, the reliability engineer tracks down the information and completes the entry correctly. The "code police" are also assessed quarterly on the accuracy of outage records in their location.

Those who police outage data make sure, to the best of their ability, that all outages in their location are reported accurately and completely. Therefore, these individuals should be rewarded for the accurate and complete reporting of outages. To avoid conflicts of interest, these individuals should not have annual outage indices such as SAIFI or SAIDI on their scorecards.

References

Blackburn, J.L., *Protective Relaying: Principles and Applications*, Marcel Dekker, New York, 1987.

California Energy Commission, Reducing wildlife interactions with electrical distribution facilities, 1999.

Chow, M.Y. and Taylor, L.S., A novel approach for distribution fault analysis, *IEEE Trans. Power Delivery*, 8(4), 1882–1889, October 1993.

Chow, M.Y. and Taylor, L.S., Analysis and prevention of animal-caused faults in power distribution systems, *IEEE Trans. Power Delivery*, 10(2), 995–1001, April 1995.

Darveniza, M., Mercer, D.R., and Watson, R.M., An assessment of the reliability of in-service gapped silicon-carbide distribution surge arresters, *IEEE Trans. Power Delivery*, 11(4), 1789–1797, October 1996.

Dugan, R.C., Ray, L.A., Sabin, D.D., Baker, G., Gilker, C., and Sundaram, A., Impact of fast tripping of utility breakers on industrial load interruptions, *IEEE Ind. Appl. Mag.*, 2(3), 55–64, May–June 1996.

EPRI 1001665, Power quality improvement methodology for wires companies, Electric Power Research Institute, Palo Alto, CA, 2003.

EPRI 1008480, Electric distribution hazard tree risk reduction strategies, Electric Power Research Institute, Palo Alto, CA, 2004.

EPRI 1008506, Power quality implications of transmission and distribution construction: tree faults and equipment issues, Electric Power Research Institute, Palo Alto, CA, 2005.

EPRI TE-114915, Mitigation of animal-caused outages for distribution lines and substations, Electric Power Research Institute, Palo Alto, California, 1999.

EPRI TR-100218, Characteristics of lightning surges on distribution lines. Second phase — final report, Electric Power Research Institute, Palo Alto, California, 1991.

Finch, K., Understanding tree outages, EEI Vegetation Managers' Meeting, Palm Springs, CA, May, 1 2001.

Finch, K., Understanding line clearance and tree-caused outages, EEI Natural Resources Workshop, April 1, 2003a.

Finch, K.E., Tree caused outages — understanding the electrical fault pathway, EEI Fall 2003 Transmission, Distribution and Metering Conference, Jersey City, NJ, 2003b.

General Electric Co. (task force of eight utility companies and the General Electric Company), Investigation and evaluation of lightning protective methods for distribution circuit. I. Model study and analysis, *IEEE Trans. Power Appar. Syst.*, PAS-88(8), 1232–1238, August 1969a.

General Electric Co. (task force of eight utility companies and the General Electric Company), Investigation and evaluation of lightning protective methods for distribution circuits. II. Application and evaluation, *IEEE Trans. Power Appar. Syst.*, PAS-88(8), 1239–1247, August 1969b.

Guggenmoos, S., Effects of tree mortality on power line security, *J. Arboriculture*, 29(4), 181–196, July 2003.

Hamilton, J.C., Johnson, R.J., Case, R.M., and Riley, M.W., Assessment of squirrel-caused power outages, in *Vertebrate Pest Control and Management Materials*, 6, K.A. Fagerstone and R.D. Curnow, Eds., Philadelphia, PA, American Society for Testing and Materials, 1989, 34–40. ASTM STP 1055.

IEEE Working Group on Distribution Protection, Distribution line protection practices industry survey results, *IEEE Trans. Power Delivery*, 10(1), 176–186, January 1995.

Johnston, L., Tweed, N.B., Ward, D.J., and Burke, J.J., An analysis of Vepco's 34.5-kV distribution feeder faults as related to through fault failures of substation transformers, *IEEE Trans. Power Appar. Syst.*, PAS-97(5), 1876–1884, 1978.

Kaempffer, F. and Wong, P.S., Design modifications lessen outage threat, *Transmission Distribution World*, October 1, 1996.

Lat, M.V. and Kortschinski, J., Distribution arrester research, *IEEE Trans. Power Appar. Syst.*, PAS-100(7), 3496–3505, July 1981.

McDaniel, J., Line arrester application field study, IEEE/PES Transmission and Distribution Conference and Exposition, 2001.

McDermott, T.E., Short, T.A., and Anderson, J.G., Lightning protection of distribution lines, *IEEE Trans. Power Delivery*, 9(1), 138–152, January 1994.

Parrish, D.E., Lightning-caused distribution circuit breaker operations, *IEEE Trans. Power Delivery*, 6(4), 1395–1401, October 1991.

Puget Sound Energy, TreeWatch Program Annual Report, 2003. http://www.wutc.wa.gov/rms2.nsf/0/e25e1114a94abbf788256dd50082d992?OpenDocument.

Short, T.A., Fuse saving and its effect on power quality, in EEI Distribution Committee Meeting, 1999.

Short, T.A., *Electric Power Distribution Handbook*, CRC Press, Boca Raton, FL, 2004.

Short, T.A. and Ammon, R.A., Instantaneous trip relay: examining its role, *Transmission Distribution World*, 49(2), 1997.

Short, T.A. and Ammon, R.H., Monitoring results of the effectiveness of surge arrester spacings on distribution line protection, *IEEE Trans. Power Delivery,* 14(3), 1142–1150, July 1999.

Simpson, P., EUA's dual approach reduces tree-caused outages, *Transmission Distribution World,* 22–18, August 1997.

Simpson, P. and Van Bossuyt, R., Tree-caused electric outages, *J. Arboriculture,* May 1996.

Taylor, L., The illusion of knowledge, Southeastern Electric Exchange Power Quality and Reliability Committee, Dallas, TX, 2003.

Tyner, J.T., Getting to the bottom of UG practices, *Transmission Distribution World,* 50(7), 44–56, July 1998.

Westinghouse Electric Corporation, Applied protective relaying, 1982.

Xu, L., Chow, M.-Y., and Taylor, L.S., Analysis of tree-caused faults in power distribution systems, 35th North American Power Symposium, University of Missouri–Rolla, October 20–21, 2003.

Yu, P., Wong, P.S., and Kaempffer, F., Tension of conductor under concentrated loads, *J. Appl. Mech.,* 62, 802–809, September 1995.

Zocholl, S.E., AC motor protection, Schweitzer Engineering Laboratories, 2003.

For Product Safety Concerns and Information please contact our EU representative GPSR@taylorandfrancis.com Taylor & Francis Verlag GmbH, Kaufingerstraße 24, 80331 München, Germany

Printed and bound by CPI Group (UK) Ltd, Croydon, CR0 4YY
01/05/2025
01858472-0001